THE
COLLINS
LONGMAN
MAPSKILLS
ATLAS

Editorial Adviser

Richard Kemp

County Adviser for Humanities, Buckinghamshire

C O N T E N T S

What are maps for?	2	
Using Symbols	3	
Compass Directions	4	
Map Grids	5	
Latitude and Longitude	6	
Using the Index	7	
Recognising Shapes	8	
Satellite Photos	9	
Scale and Measuring Distance	10-11	

BRITISH ISLES AND EUROPE

British Isles	Relief	12-13
British Isles	Countries, Counties and Regions	14-15
British Isles	Population	16-17
British Isles	Climate	18-19
England and Wales		20-21
Scotland		22
Ireland		23
British Isles	Early Times	24
British Isles	Middle Ages and 16th and 17th Century Britain	25
British Isles	The Agricultural and Industrial Revolutions	26
British Isles	People on the move in 19th and 20th centuries	27
British Isles	Work	28-29
British Isles	Sport and Leisure	30-31
Europe	Relief	32
Europe	Countries	33
Europe	Climate and Holidays	34-35
Europe	Environment	36-37
Europe	Standard of Living and People on the move	38-39

REST OF THE WORLD

Africa	Relief	40
Africa	Countries	41
Africa	Vegetation	42-43
Asia	Relief	44
Asia	Countries	45
Australasia	Relief	46
Australasia	Countries	47
North America	Relief	48
North America	Countries	49
South America	Relief	50
South America	Countries	51
World	Relief	52-53
World	Political	54-55
World	Climate	56-57
World	Population	58-59
World	Religions	60-61
World	Rich and Poor	62-63
World	Natural Disasters	64-65
World	Exploration	66-67
Solar System and Space Exploration		68-69
Index		70-73

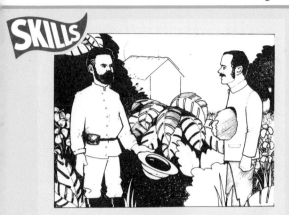

"Dr Livingstone I presume" is what Stanley is supposed to have said when he finally found him in the middle of Africa in 1871. Finding him had been an amazing achievement. Livingstone had no communication with the outside world – there was no post and telephones and radio had not yet been invented. Above all Stanley had the problem that he had no **maps** of the vast area of Africa in which he was looking for a single person.

Only a hundred years ago there were still plenty of places in the world that were either unknown, or known only to a very few people. The maps of these places, if they existed at all, were not very helpful – lots of blank spaces or vague, inaccurate information.

Nowadays there is not a single corner of the world that we cannot photograph in great detail. The technology of **satellite photography** has reached a stage when a photo taken from space can pick out individual people on the earth's surface!

If satellite photos, and **aerial photos** taken from planes, are so detailed, why do we still need to draw maps? The answer is that photos and maps are both useful, but that they tell us different sorts of information.

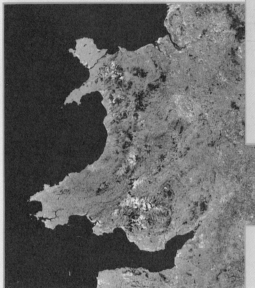

▲ **SATELLITE PHOTO**

One important difference between a photo and a map is this: a photo has to show everything there is to see at the moment the camera shutter clicks. Some of the information can be confusing. When drawing a map we can **select** only the information we want to show about an area. In this way a map can **simplify** by cutting out information we do not need.

▲ **MAP A**
WALES Relief

▲ **MAP B**
WALES Counties

Look carefully at the satellite photo and the two maps of Wales.

1 What sort of information has been selected for the Relief Map, **A**?

2 What sort of information has been selected for the Counties Map **B**?

3 Choose one or two large and obvious features from each map.
• Try and pick out those features on the satellite photo.
• For your chosen features, which is more useful, the map or the photo?

4 For what sort of things or uses might a satellite photo be better than a map?

5 ATLAS SKILLS
What different **types** of maps are there in this atlas?
• Look through the atlas, and make a list of the different types of maps – **Relief** and **Counties** are just two of the different map titles.
• For each different type of map make a list of the sort of information the map shows.

◀ **AERIAL PHOTO**

MAP
▼

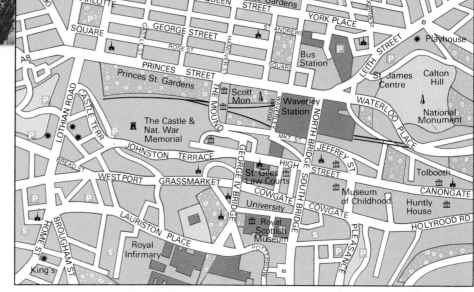

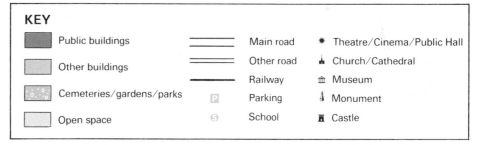

KEY

▨ Public buildings		Main road	✳	Theatre/Cinema/Public Hall
▢ Other buildings		Other road	⚓	Church/Cathedral
▨ Cemeteries/gardens/parks	Ⓟ	Parking	🏛	Museum
▢ Open space	Ⓢ	School	⚱	Monument
			🏰	Castle

One of the ways a map **selects** and **simplifies** information is by using **symbols**.

You can see how this works by noticing the differences between the aerial photo and map. Both show the same area in Edinburgh.

What are the different sorts of symbols?

• **Colours**
• **Drawings and shapes**
• **Words, letters and numbers**

The symbols used on this map are shown in the map **key**. The key tells you what each symbol **represents**, or stands for.

Look carefully at the map:

1 Write about three ways that the map uses different colours to tell us information about the area.

2 Write about four ways that the map uses drawings or shape symbols – say what each symbol looks like, and what it represents.

3 Look at the way the map uses words and letters. How many different ways are these sorts of symbols used on the map?

4 *Look at the aerial photo, it can give us some information that the map cannot.*
• What is most of the land used for?
• If you were going to live somewhere in this area, where would you choose? Why?

5 **ATLAS SKILLS**
Look at some other maps in this atlas - for example the maps on pages 32 and 33.
• What colour symbol is used for mountain peaks?
• How can you tell the difference between the name of a country, and the name of a town or city?
• How can you tell which is the capital city of a country by looking at just the symbols?

One of the things we use maps for is to find out where places are. Often we want to know where one place is in relation to another place. For example, where is Scotland in relation to England?

SCOTLAND IS ABOVE ENGLAND

One thing is for sure, Scotland is not above England!
What we need is an accurate way of giving the **direction** from one place to another. To do this we use **compass directions.**

Wherever you are in the world the needle of a compass always points in one direction – towards the **north.** Because this north direction never changes, we can use it to fix other directions as well. These directions are called **compass points**, or **compass directions.**

THE FOUR POINT COMPASS

These are the four main compass points.

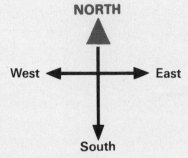

Naughty **E**lephants **S**quirt **W**ater is one way to remember them!

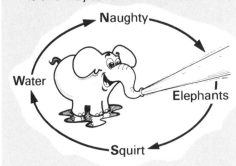

THE EIGHT POINT COMPASS

We can make more compass points by dividing the four main ones – like this:

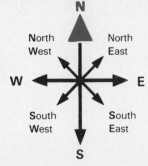

THE SIXTEEN POINT COMPASS

You have seen how the four main points of the compass – North, East, South and West – can be divided to make an eight point compass. These eight points can be divided once more, to make a sixteen point compass. The way this is done is shown in the diagram below.

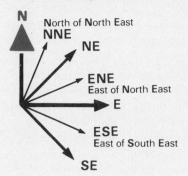

Use this information to draw your own sixteen point compass.

1 On a piece of tracing paper, draw your own eight point compass.

2 The map on this page shows some of the main cities in Britain.

• Make a copy of the table.

• Use your compass to work out the compass directions that have been left blank in the third column. Fill in the correct directions on your copy of the table.

FROM	TO	COMPASS DIRECTION
BIRMINGHAM	BRADFORD	?
	BOURNEMOUTH	?
	BELFAST	?
LONDON	BIRMINGHAM	?
	CARDIFF	?
	NORWICH	?
GLASGOW	EDINBURGH	?
	BELFAST	?
	ABERDEEN	?

3 **ATLAS SKILLS**
Look at the map of the world on page 54-55.
• Is North America east or west of Europe?
• Is the Indian Ocean east or west of the Atlantic Ocean?
• Starting from London, in which direction would you have to fly to get to:

AUSTRALIA	BRAZIL
CANADA	CHINA

We use maps to find and locate things – buildings, places, countries, oceans, or whatever we are looking for.

When we have found the things we are looking for, we need a good way of **recording** exactly where they are. The normal way to do this is by using a **map grid.**

Look at the aerial photo. It has a grid of squares over it – notice that all the grid squares are the same size.

We can locate any one grid square by using the letters and numbers, which are printed along the bottom and side of the photo.

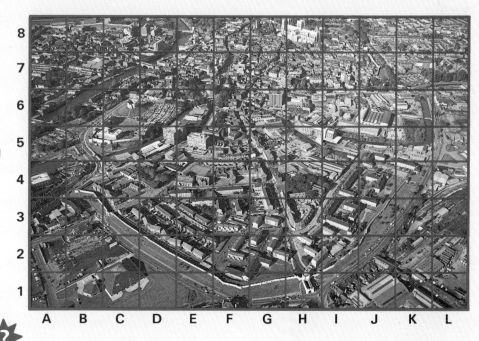

- The square in the bottom left corner is **A1.** A1 is the **grid reference** for this square.

- Find York Minster which is near the top of the photo. The grid reference for the square which contains York Minster is **H8.**

1 *Study the photo carefully:*

- There are a number of bridges in the area covered by the photo.
  How many can you find? Write down the grid reference for each bridge you find.

- Write down the grid reference for two churches and three car parks.
- What do you think that the large building in square **C6** is used for? Explain your answer fully.

The grid you used with the aerial photo gave you the grid reference for a **square**. The grid on the map of Eagle Island will give grid references for particular **points** on the map. Look at the Eagle Island map grid – how is it different from the photo grid?

- It uses only numbers.

- It is each *line* that is given a number, not each square

How to use the map grid:
- Find the fort near the middle of Eagle Island. It is located where two grid lines cross.
  - look along the bottom of the map – the grid line is numbered **15**
  - now look up the side – the grid line is numbered **34**
  - put those two grid numbers to-gether, **15 34**, and you have the grid reference for the fort.

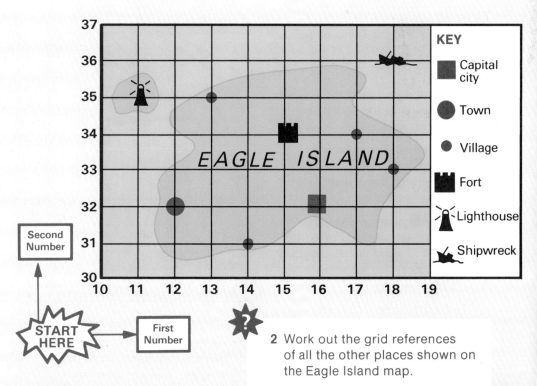

2 Work out the grid references of all the other places shown on the Eagle Island map.

When Columbus and his crew sailed across the Atlantic, many people at the time thought that he would sail right off the edge of the world - they thought that the earth was flat!

The achievement of those early explorers was all the greater because they journeyed without maps, and without a proper system of working out where on earth they were.

Today we have a system for working out the exact location of any place on the earth's surface. The system uses **lines of latitude** and **lines of longitude,** and works very much like the grid lines on a map.

To make a grid system, two sets of lines are drawn around the globe.
- the lines drawn in blue are called **lines of latitude**
- the lines drawn in red are called **lines of longitude**

We usually find it more convenient to look at the earth on a flat map, rather than on a globe. This map of the world shows the same grid, made up of lines of latitude and lines of longitude.

Unlike some people in the times of those early explorers, we know that the earth is a globe.

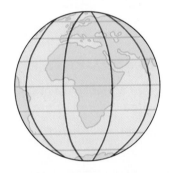

The continents are coloured green and the sea pale blue.

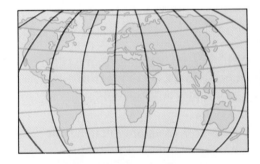

To be able to use a grid to locate places, the grid lines must have numbers or letters.

- The most important line of latitude is the **equator,** the equator is given the number **0.**

- All the other lines of latitude are given numbers, based on how far they are from the equator.This is explained on page 7.

- Lines of latitude **north** of the equator have a number *and* the letter **N.** Lines of latitude **south** of the equator have a number *and* the letter **S.**

- The most important line of longitude is the one that is given the number **0.** This line passes through the Greenwich Observatory, near London, and it is called the **Greenwich Meridian.**

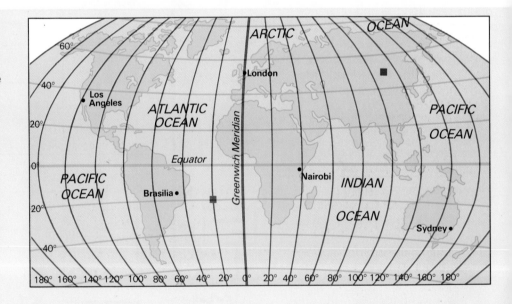

- All the other lines of longitude are given numbers, based on how far they are from the Greenwich Meridian (GM).

- Lines of longitude **west** of the GM have a number *and* the letter **W.** Lines of longitude **east** of the GM have a number *and* the letter **E.**

MEASURING LATITUDE AND LONGITUDE

The number given to a line of latitude is worked out by measuring the angle that line makes with the line of the equator – as the diagram shows:

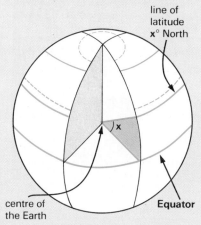

x° is the 'angle' between the Equator and the line of latitude

The number given to a line of longitude is worked out by measuring the angle that line makes with the Greenwich Meridian - as the diagram shows

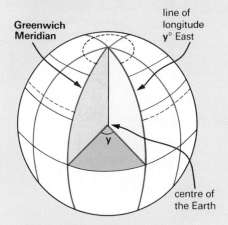

y° is the 'angle' between the Greenwich Meridian and the line of longitude

1 What is the highest number a line of latitude can have? Which two places on earth have that latitude?

2 What is the highest number a line of longitude can have?

3 Which three lines of latitude or longitude are the only one that do not need a letter as well as a number? Why don't they?

To pinpoint a place, you have to use a latitude reference and a longitude reference together. Look at the world map. Some lines of latitude and longitude are shown and numbered. The map cannot show every single line, so it gives you lines at intervals of 20, as a guide.

- *Find the red square in the Atlantic Ocean:*
  - it's latitude reference is **20S**
  - it's longitude reference is **20W**
  - so the exact location is written as **20S 20W**.

- *Find the purple square in Asia:*
  - it's latitude reference is **50N**
  - it's longitude reference is **110E**
  - so its exact location is written as **50N 110E**.

1 Which continents are found at each of these references?

20S 60W   50N 110W   27S 133E
20N 20E   35N 90E

2 Five cities are shown on the map. Write down the reference of each one.

3 ATLAS SKILLS
Look at the World Countries map on page 54-55. Each of these references locates a country. Collect the first letter of each name, and you will have the letters to spell a word – unjumble the letters to make the word.

40S 70W   25N 18E   8N 81E
5S 35E   32N 66E

USING THE INDEX

Most places that are named in the atlas are listed in the index, which is at the back of the atlas.

Each entry in the index is organised in a similar way. This is an enlarged version of an index entry, which explains what information the index gives you.

place name	page number	latitude

New York   U.S.A.   49   **B5**   34N   118W

	country name	grid reference	longitude

1 At the front of the index, on page 70, there is a list of the abbreviations used in the index. What do these abbreviations mean?

mtn.   des.   r.   l.

2 How many places listed in the index have the word **Newtown** in their name? How many different countries have a town called **Newtown**.

3 Using the index find out where in the world these interesting sounding places or features are! Say whether each one is a town, island, lake, river or mountain.

Bug	**Great Bear**
Rhum	**Ouagadougou**
Ob	**Macgillycuddy's Reeks**
Darling	**Orinoco**

This is an old map of the world, drawn by a man called Hondius who lived in Flanders 350 years ago. It is not easy to recognise some of the bits of his map! Yet, considering the little information he had to go on, it is a remarkable effort.

Nowadays we know exactly what the continents and seas look like. The shape of, say, Africa on one modern map is the same as the shape of Africa in all the other maps.

On these pages are some tasks which will give you the chance to test your skill at recognising shapes on maps.

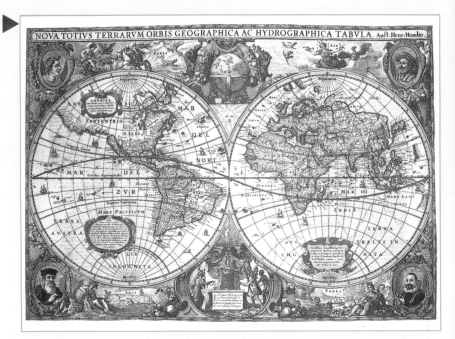

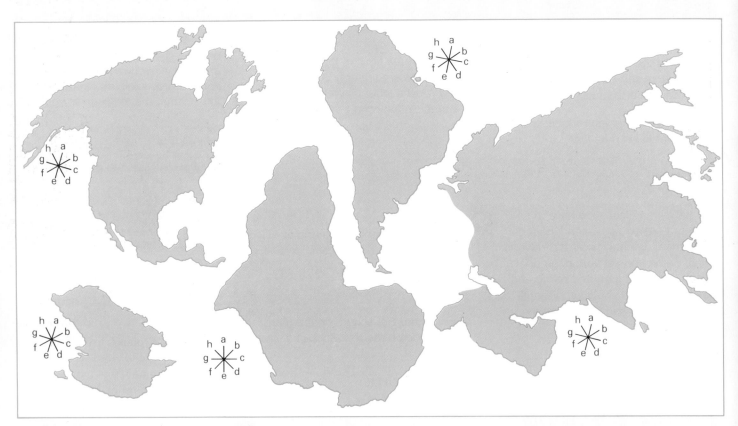

CONTINENTAL CONFUSION

Here are maps showing the outline of all or part of five different continents – the problem is that they are not all the right way up!

1 Name each of the continents shown here. If you need help you can compare these shapes with the world map on pages 52-53, which shows all the continents the right way up.

2 Each of the maps shown here has a compass rose next to it – the problem is that the compass directions have not been filled in. Taking each map in turn, say which of the letters **a** to **h** represents the compass direction of **North** for the map.

ISLAND SURPRISE

Here are outline maps which show the shape of some of the world's larger islands. They are not all the right way up! But, each map has a compass rose showing the direction of north.

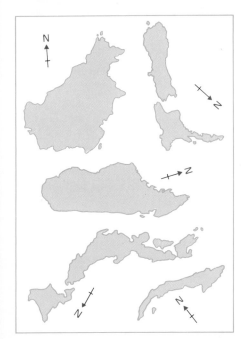

Find out the names of all the islands. Do this by comparing their shapes in these outline maps with the world map on pages 52-53.

RIVER RECOGNITION

These maps show the shape of four of the world's longest rivers.

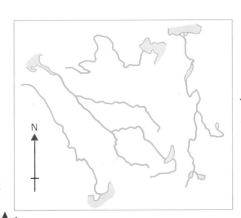

To find out the names of these rivers, compare their shapes with the river shapes you can see on the world map on pages 52-53.

Satellite photos are now used for a wide range of things. You probably see satellite photos almost every day, when you watch the weather forecast. These photos come from Meteosat, a weather satellite which hovers over the coast of West Africa.

Other satellite photos provide information about forestry and crops, mineral resources and rock formations, ocean currents, pollution and many other things.

Most photos from satellites are taken and stored electronically. They are then sent back to earth as radio waves, which are decoded at a ground station. Using computers, scientists can alter the colours used in the photos. This can be useful when colour coding particular features.

MAP

The most detailed photos from space are taken using cameras with photographic film. A satellite camera, using a telephoto lens, can pick out a single person, and tell us what colour shoes he or she is wearing! The exposed film is ejected at intervals and parachuted back to the ground.

This is a photo of the city of **San Francisco** in the USA. The photo was taken from a space satellite orbiting above the earth's surface.

SATELLITE PHOTO

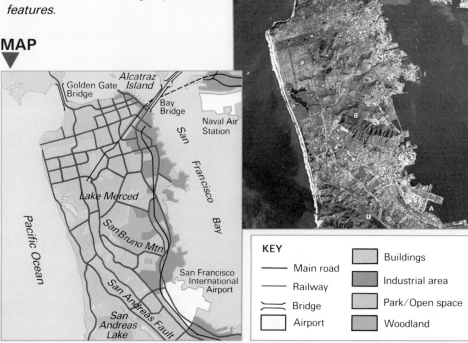

KEY

—— Main road	Buildings
— Railway	Industrial area
⟩⟩ Bridge	Park/Open space
☐ Airport	Woodland

Compare the photo with the map showing the same area.

1 Which of the letters printed on the photo match each of these features?

  San Andreas Fault
  San Francisco Bay
  Airport
  Alcatraz Island

2 Two bridges are shown by letters on the photo. Which one is the Golden Gate?

3 What does the photo tells us about the layout of streets in much of San Francisco?

4 What is the land being used for at the places marked B, F and H on the photo?

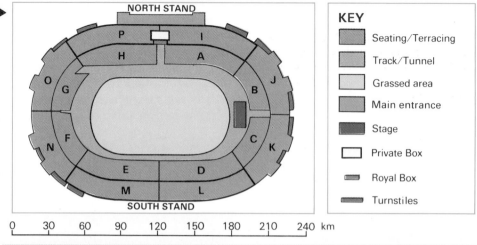

MAP 1 ▶

One of the things we use maps for is to work out how far it is from one place to another. We can do this because most maps are carefully and accurately drawn **to scale**.

Anything drawn on a map – a building, road, Wembley Stadium or whatever – is smaller on the map than it is in real life. What the **map scale** tells us is just how much smaller things have been drawn on the map.

KEY

- Seating/Terracing
- Track/Tunnel
- Grassed area
- Main entrance
- Stage
- Private Box
- Royal Box
- Turnstiles

Scale: 0 30 60 90 120 150 180 210 240 km

The photo shows a concert at Wembley Stadium. Wembley is a big place, some people are going to be a long way from the stage. Using the map of the stadium, and its scale, we can work out the actual distances.

HOW TO USE THE SCALE

- The line below the map of the stadium is the map scale.
- The scale line is divided into sections, each one centimetre long.
- The scale tells us that each centimetre on the map **represents** 30 metres in the actual stadium.

Question: if you had a seat in the middle of Section P, how far would you be from the stage?

To work out the answer:

1 Locate Section P and the stage on the stadium map.

2 Using your ruler, or the edge of a piece of paper, measure the distance, on the map, between the two points.

3 Place your ruler (or paper edge) against the map scale, and read off what the distance would be actually on the ground in the stadium.

Now work out the actual distances between these points:

Section K and the Royal Box

The Stage and Section F

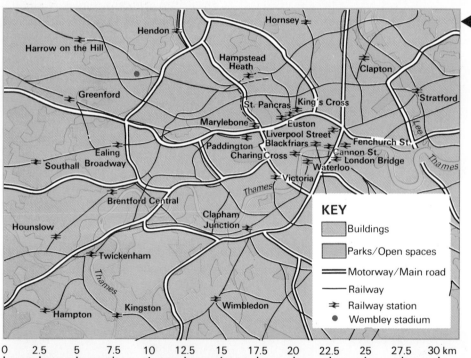

MAP 2 ◀

Here is another map, **Map 2**, which shows Wembley Stadium. But, this map shows much more than just the stadium, as it covers most of London.

The information on this map has been drawn at a different scale from the Wembley Stadium map. The stadium is too small to be shown as anything more than a dot.

Look at the scale line for this map. It shows that every centimetre on the map represents 2.5 kilometres actually on the ground.

Although the scale of this map is different, you use the scale line in exactly the same way as you did for the Wembley Stadium map.

KEY

- Buildings
- Parks/Open spaces
- Motorway/Main road
- Railway
- Railway station
- Wembley stadium

Scale: 0 2.5 5 7.5 10 12.5 15 17.5 20 22.5 25 27.5 30 km

Here is another map, **Map 3**, drawn at another different scale. This map includes London, and shows where London is located compared to some other towns.

This map has been drawn at a much smaller scale than **Map 2**. The scale is now too small for the map to even show where Wembley Stadium is within London.

MAP 3

▼

Scale 1: 10 000 000

0   100   200   300   400   500 km

DIFFERENT TYPES OF SCALES

The maps in this atlas have a **line scale,** like the ones you have been using. The proper name for this type of scale is a **linear scale.**

On **Map 3** the scale is also shown another way. This method is called **a representative fraction scale.**

Look at the scale of the Map 3.

Next to the line scale is written 1: 10 000 000 . This means that one centimetre on the map represents 10 000 000 centimetres actually on the ground.

LARGER AND SMALLER SCALES

As a map is drawn to cover a **larger area,** it has to use a **smaller scale.**

You can use the representative fractions to compare maps at different scales. Compare **Map 2** and **Map 3** for example.

Map 2 has a scale of 1: 250 000,

Map 3 has a much smaller scale 1: 10 000 000,

– divide 10 000 000 by 250 000

– 1000 ÷ 25 = 40

– so Map 3 has a scale 40 times smaller than Map 2.

1 You have arrived at Kings Cross railway station, and want to go to Wembley Stadium. How far is it between the two places?

2 Which of the railway stations shown on the map is nearest to the stadium? What is the distance?

3 If you were really travelling from Kings Cross to Wembley Stadium, you would have to cover a longer distance than the one you have just measured. Explain why this would be the case.

4 *You are planning a tour going round all the towns shown on Map 3. You can visit the towns in any order but your aim is to visit all the places by travelling the shortest possible distance.*

• What do you think would be the shortest route to visit all the towns. List the towns in the order in which you would visit them.

• Use the scale to work out the actual distance between each of the towns on your tour.

• What would be the total distance you would travel to visit all the towns.

5 **ATLAS SKILLS**
Turn to the world map on pages 54-55. This map has the smallest scale of all the maps in the atlas, as it covers the largest area. Using the scale work out the distance if you were flying from London to each of these world cities.
NEW YORK, USA

RIO DE JANEIRO, Brazil

MOSCOW, USSR

NAIROBI, Kenya

Which capital city in the world is furthest from London? How far away is it?

PHOTO A ◀

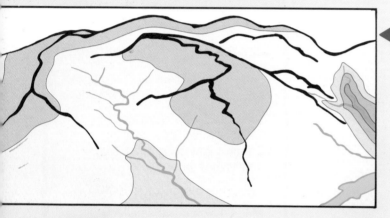

DIAGRAM B ◀

MAP C ◀

▲ Ben Nevis
1344

1223
▲ Carn
Mór
Dearg

KEY

The colours show the
height of the land
above sea level

Over 1000 metres
500-1000 metres
200-500 metres
100-200 metres
0-100 metres

▲   Mountain peak

Very few places on the earth are
completely flat. One of the things we
use maps for is to show information
about the shape of the land. We use
the word **relief** when we talk about
the shape of the land in a particular
area.

A mountainous area has **high relief**.
A generally flat area has **low relief**.
An area with hills and valleys has
undulating relief.

Photo A shows the mountains
around Ben Nevis in Scotland. It is
clearly an area of high relief. The
photo gives us an idea of the shape
of the land, but it does not tell us the
height (or **altitude**) of different parts
of area.

Diagram B is a drawing of the
photo. The drawing uses a way of
showing how high the land is.
Places at different heights are
shaded in different colours. This
method is called **layer colouring**.

The photo and drawing show a side,
or oblique aerial view, of this part of
the mountains. **Map C** shows what
the area is like if you were looking
down on it from directly above. The
map uses the same method of layer
colouring as the drawing.

1   *Look at the key on Map C:*
    • How many different layer
      colours are shown in the key?
    • Are all these layer colours
      actually used on Map C?

2   The key tells you that the
    highest ground on Map C is
    above what height?

3 · How does Map C show the
    location of the highest
    mountain peaks? What is the
    name and actual height of the
    highest peak?

    Look at the relief map of Britain
    opposite. It uses the same layer
    colouring to show the relief of
    different parts of Britain. It tells us
    the height of land above **sea level**.

4   Which are the highest mountain
    peaks in Scotland, Wales,
    England and Ireland? What are
    their heights? In which mountain
    ranges are each of these
    mountains found?

5   Name two areas of Britain that
    have mostly low relief.

13

The colours show the height of the land above sea level

- Over 1000 metres
- 500-1000 metres
- 200-500 metres
- 100-200 metres
- 0-100 metres

▲ 123 Mountain peak (height in metres)

〜 River

Lake

Scale 1:4 000 000

0   50   100   150 km

1 centimetre on the map represents
40 kilometres on the ground

© Collins ♦ Longman Atlases

Shetland Islands
Yell, Unst, Mainland, Whalsay, Foula, Bressay, Sumburgh Head, Fair Isle

Orkney Islands
Westray, Stronsay, Mainland, Hoy, Scapa Flow, Pentland Firth

ATLANTIC OCEAN

St. Kilda

Outer Hebrides
Lewis, Butt of Lewis, Cape Wrath
Harris
North Uist
South Uist
Barra

Inner Hebrides
Skye, Cuillin Hills 1009
Rhum
Coll
Tiree
Mull
Ben More 966
Jura
Islay

The Minch
North West Highlands
Loch Maree
Loch Shin
Dornoch Firth
Ben Wyvis 1045
Moray Firth
Kinnaird's Head
Loch Ness
Spey
Don
Cairngorms
Ben Macdhui 1311
Grampian Mountains
Dee
South Esk
Ben Nevis 1343
Ben Lawers 1214
Loch Tay
Tay
Earn
Loch Awe
Loch Lomond
Ochil Hills
Firth of Tay
Firth of Lorn
Forth
Firth of Forth
Goat Fell 874
Arran
Firth of Clyde
Clyde
Southern Uplands
Tweed
Holy I.
843
Nith
Annan
Teviot
The Cheviot Hills
Tyne

Malin Head
Rathlin I.
Lough Foyle
Mts. of Antrim
North Channel
Erris Head
Donegal Bay
Mourne
Lough Neagh
Lower Lough Erne
Upper Lough Erne
Strangford Lough
Mourne Mts. 852
Dundalk Bay
Luce Bay
Solway Firth
Isle of Man
Lake District 978
Scafell Pike
Cumbrian Mts.
Cross Fell 893
The Pennines
Wear
Tees
Cleveland Hills
North York Moors
Flamborough Head

NORTH SEA

Achill I.
Lough Conn
Lough Mask
Lough Corrib
Galway Bay
Aran Is.
Lough Ree
Lough Derg
Shannon
Boyne
Shannon

IRISH SEA

Morecambe Bay
Ribble
Ouse
Aire
Humber
Spurn Head
Mersey
636 High Peak

The Wash

Nore
Barrow
Suir
Slaney
Wicklow Mts. 926
Anglesey
Caernarfon Bay
Snowdon 1085
Dee
Trent
The Fens
Norfolk Broads
Great Ouse
Nene
Macgillycuddy's Reeks 1041
Carrauntoohil
Blackwater
Lee
Galty Mts.
Dingle Bay
Cape Clear

St. George's Channel
Carnsore Point
St. David's Head
Cardigan Bay
Cambrian Mountains
Teifi
Tywi
Brecon Beacons
Wye
Usk
Severn
Cotswold Hills
Avon
Chiltern Hills
Thames
Isle of Sheppey
North Downs
Dungeness
Beachy Head

Carmarthen Bay
Bristol Channel
Exmoor
Mendip Hills
Salisbury Plain
New Forest
South Downs
Parrett
Torridge
Exe
Tamar
Bodmin Moor
Dartmoor
Stour
Solent
Isle of Wight
Lyme Bay
Bill of Portland

English Channel

Isles of Scilly
Land's End

The map, on the opposite page, shows how Britain is divided into smaller areas for local government. These areas are called **counties** or **regions**.

1 a Which county in England has the largest area?

b Which is the smallest Welsh county?

2 Find your home county or region – name the other counties or regions which border it.

LINES

Maps use **lines** to show the borders between two countries. These are called **international boundaries**. Sometimes maps also show the borders between regions inside a country. These are called **internal boundaries**. The usual rule is this: *the thicker the line on the map, the more important the boundary.*

3 a Which is the most important boundary line on the map opposite?

b Why is this boundary more important than the one between England and Scotland? (Check with the box below)

LETTERS

Maps use different types of **letters** for different types of information.

- **CAPITAL LETTERS** are used for the names of **COUNTRIES**.

- **Small letters** are used for the names of **towns and cities**.

- *Sloping, italic letters* are used for the names of *physical features* - such as *rivers, mountains, lakes*

ATLAS SKILLS

4 *Look at the map on page 33. Find the country called AUSTRIA.*

a What is the capital of Austria?
b List all the other countries which border onto Austria.

5 *Look at the map on page 32. Notice how the country boundaries are shown on this relief map. Find Austria again.*
a Give the name of the *MOUNTAINS* that cover part of Austria.
b Name one *river* that flows through Austria.

SYMBOLS

All maps show towns and cities with a **symbol**. The rule is this: *never name a town on a map without using a symbol to mark exactly where it is..*

ATLAS SKILLS

6 *Look at one of the maps on pages 20-23. Check with the map key:* How many different symbols does the map use for towns and cities? Why are different symbols used?

7 Find your home town on the map – if it is not shown, find the nearest place that is named. Which three 'cities' are nearest to where you live?

SCOTLAND

NORTHERN IRELAND

REPUBLIC OF IRELAND

ISLE OF MAN

WALES   ENGLAND

CHANNEL ISLANDS

This map shows the countries that make up the British Isles. As on all the maps of countries in this atlas, each separate country is shown in different colour.

When drawing and colouring maps of countries the rule is this; *never use the same colour for countries that are next to each other.*

USING THE RIGHT NAMES

Most people just talk about **Britain**. But at times it can be rather confusing. When British teams take part in sporting events other names are used on the results board - initials such as **UK** or **GB**.

To set the record straight these are the correct names:

British Isles includes *England, Wales, Scotland, both parts of Ireland, Isle of Man, Channel Islands*

Most of the parts of the British Isles form one main country – the **United Kingdom**. The exception is the **Republic of Ireland**, which is a completely separate country.

United Kingdom includes *England, Wales, Scotland, Northern Ireland, Isle of Man, Channel Islands*

Great Britain includes *England, Wales, Scotland*

© Collins ◇ Longman Atlases

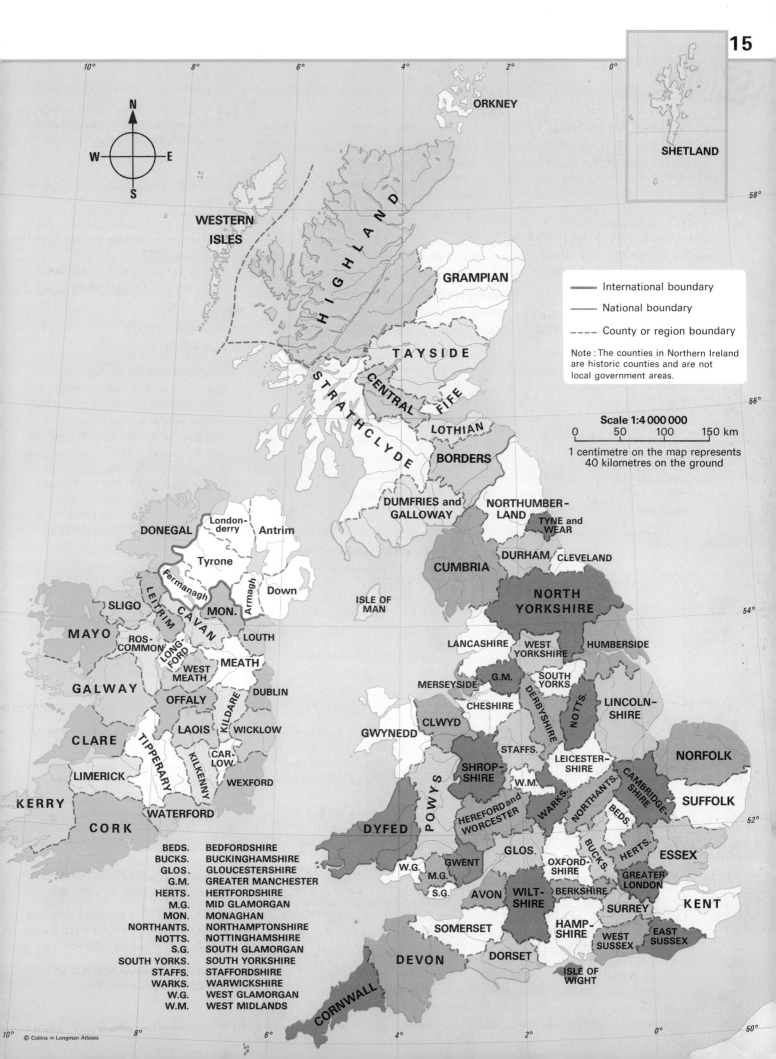

ORKNEY

SHETLAND

N
W E
S

58°

WESTERN
ISLES

HIGHLAND

GRAMPIAN

International boundary

National boundary

County or region boundary

Note: The counties in Northern Ireland
are historic counties and are not
local government areas.

TAYSIDE

CENTRAL FIFE

STRATHCLYDE

LOTHIAN

BORDERS

56°

Scale 1:4 000 000

0   50   100   150 km

1 centimetre on the map represents
40 kilometres on the ground

DUMFRIES and
GALLOWAY

NORTHUMBER-
LAND

TYNE and
WEAR

DONEGAL

London-
derry

Antrim

CUMBRIA

DURHAM

CLEVELAND

Tyrone

NORTH
YORKSHIRE

54°

SLIGO

Fermanagh

CAVAN

MON.

Armagh

Down

ISLE OF
MAN

LANCASHIRE

WEST
YORKSHIRE

HUMBERSIDE

MAYO

LEITRIM

ROS-
COMMON

LONG-
FORD

LOUTH

MERSEYSIDE

G.M.

SOUTH
YORKS

WEST
MEATH

MEATH

CHESHIRE

DERBYSHIRE

NOTTS.

LINCOLN-
SHIRE

GALWAY

OFFALY

DUBLIN

CLWYD

STAFFS.

LEICESTER-
SHIRE

NORFOLK

LAOIS

KILDARE

WICKLOW

GWYNEDD

CLARE

TIPPERARY

CAR-
LOW

SHROP-
SHIRE

W.M.

CAMBRIDGE-
SHIRE

SUFFOLK

52°

LIMERICK

KILKENNY

WEXFORD

P
O
W
Y
S

HEREFORD and
WORCESTER

WARKS.

NORTHANTS.

BEDS.

HERTS.

ESSEX

KERRY

WATERFORD

DYFED

GLOS.

OXFORD-
SHIRE

BUCKS.

GREATER
LONDON

CORK

W.G.

GWENT

BERKSHIRE

KENT

M.G.

AVON

WILT-
SHIRE

SURREY

S.G.

HAMP-
SHIRE

EAST
SUSSEX

BEDS.   BEDFORDSHIRE
BUCKS.   BUCKINGHAMSHIRE
GLOS.   GLOUCESTERSHIRE
G.M.   GREATER MANCHESTER
HERTS.   HERTFORDSHIRE
M.G.   MID GLAMORGAN
MON.   MONAGHAN
NORTHANTS. NORTHAMPTONSHIRE
NOTTS.   NOTTINGHAMSHIRE
S.G.   SOUTH GLAMORGAN
SOUTH YORKS. SOUTH YORKSHIRE
STAFFS.   STAFFORDSHIRE
WARKS.   WARWICKSHIRE
W.G.   WEST GLAMORGAN
W.M.   WEST MIDLANDS

SOMERSET

DORSET

WEST
SUSSEX

DEVON

ISLE OF
WIGHT

50°

CORNWALL

© Collins ◇ Longman Atlases

10°   8°   6°   4°   2°   0°

POPULATION DENSITY

The *population* of a place is the total number of people who live there. In urban areas, towns and cities, there is a large population living in a small area – we can say that there is a high population **density**. Population density in urban areas is usually over 150 people living in each square kilometre of land.

In the countryside, rural areas, the population density is lower –between 10-150 per sq km. Very few people live in the most mountainous places – the population density is under 10 per sq km.

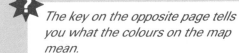

The key on the opposite page tells you what the colours on the map mean.

1 What is the population density in your home area?

2 Which parts of Britain have the highest and lowest population densities?

POPULATION GROWTH

This **line graph** shows how the population of Britain has changed since 1801 – the line at the top of the graph is the one to look at. The line is always going upwards, so this means that Britain's population has been growing larger since 1801. The steeper the line on the graph, the faster the population growth.

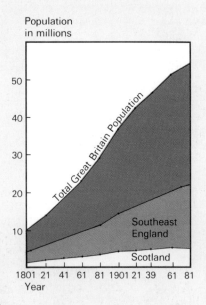

Population in millions

POPULATION PYRAMIDS

The **population pyramids** show the number of people of different ages in the total population – younger generations at the bottom, older generations at the top.

England and Wales 1861

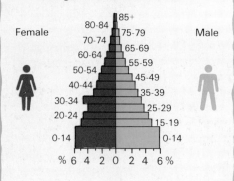

England and Wales 1981

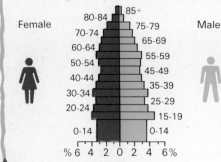

POPULATION DISTRIBUTION

Just over 55 million people live in the U.K. If you look at the map opposite you can see that those 55 million people are very unevenly spread over the country. The pattern of how the population is spread over a country is called its population **distribution**.

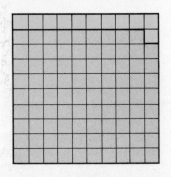

Each square represents 1% of the population

☐ Urban population   ☐ Rural population

8 *Using the graph above, work out:*
   a What percentage of Britain's population lives in rural areas?

   b What is the urban percentage?

3 Using the graph, work out what the population of Britain was in 1901 and 1981.

4 In which 50 year period, since 1801, was Britain's population growing the fastest?

5 How has the population of Scotland and Southeast England changed since 1801?

6 What other information does each population pyramid show?

7 How are the shapes of the pyramids for 1861 and 1981 different? What does this tell us about the numbers of older and younger people at those two dates?

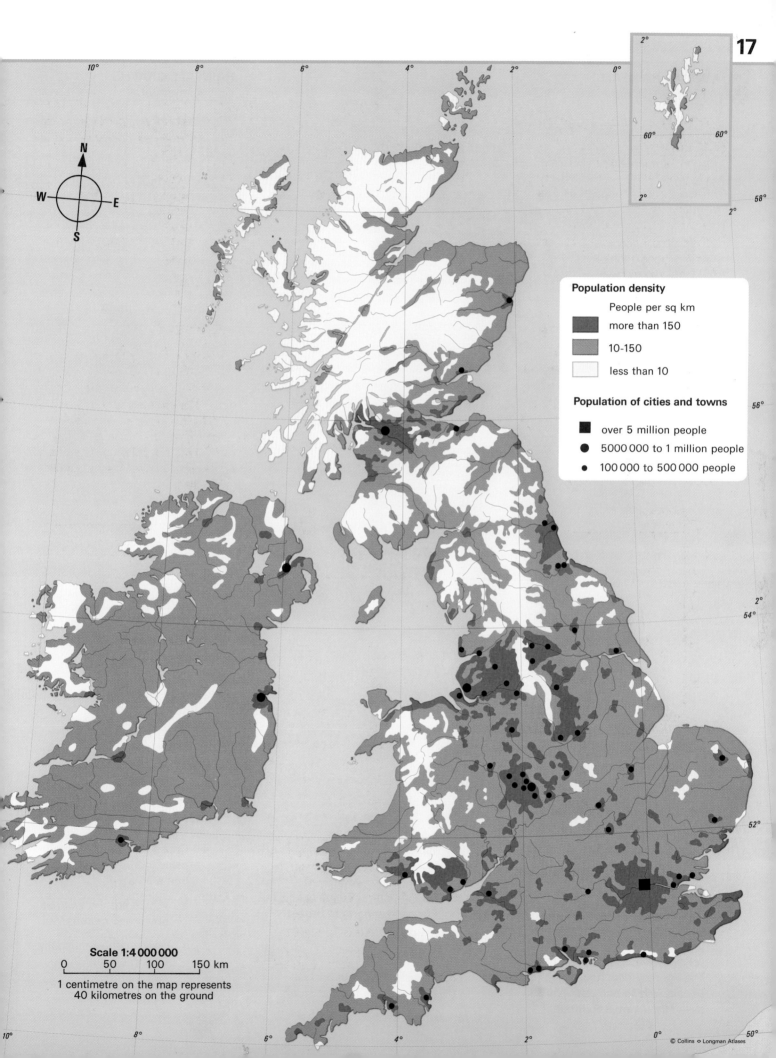

Population density

People per sq km

more than 150

10-150

less than 10

Population of cities and towns

■ over 5 million people

● 5 000 000 to 1 million people

• 100 000 to 500 000 people

Scale 1:4 000 000

0   50   100   150 km

1 centimetre on the map represents
40 kilometres on the ground

© Collins ◇ Longman Atlases

ANNUAL RAINFALL

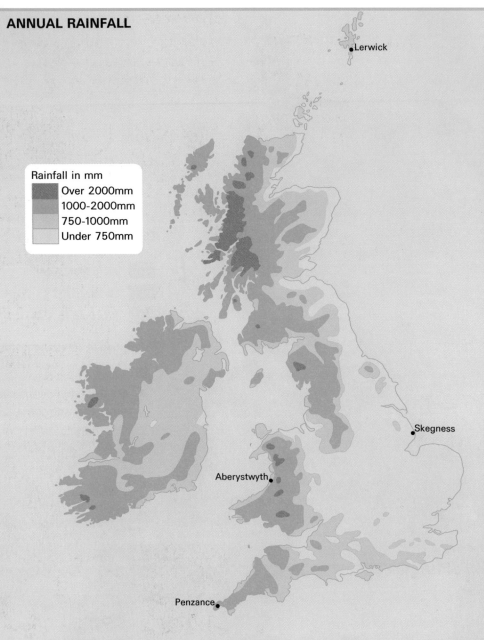

Rainfall in mm
- Over 2000mm
- 1000-2000mm
- 750-1000mm
- Under 750mm

RAINFALL GRAPHS

The bar graphs show how much rain usually falls in each month at four places in Britain.

People in Britain always seem to be complaining about how wet the weather is. The map, above, shows that some places are wetter than others.

As the map key explains, the places coloured dark green have the most rain. The area coloured yellow is the driest part of Britain. The figures show how much rain usually falls in one year - this is called the **Annual Rainfall**.

1 Which are the wettest parts of Britain?

2 How much rain usually falls in the wettest parts of Britain? Write your answer in mm (millimetres)

3 What is the usual annual rainfall in your home area?

4 Choose the graph of the place that is closest to your home area:

   a Which are the wettest and driest months?

   b How much rain usually falls in each of those two months?

TEMPERATURE

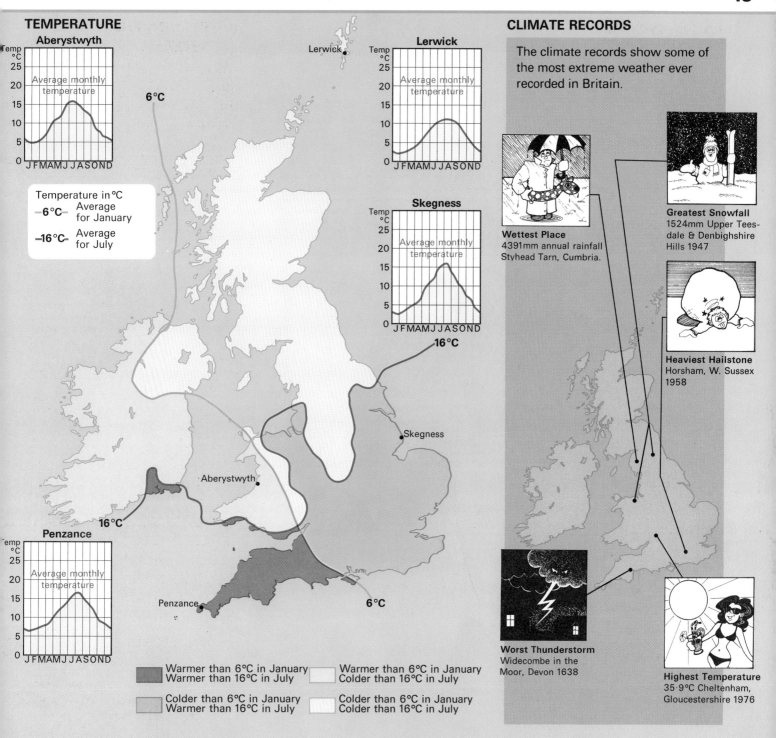

Aberystwyth

Temp °C
25 20 15 10 5 0
Average monthly temperature
J F M A M J J A S O N D

6°C

Lerwick

Temp °C
25 20 15 10 5 0
Average monthly temperature
J F M A M J J A S O N D

Lerwick

Skegness

Temp °C
25 20 15 10 5 0
Average monthly temperature
J F M A M J J A S O N D

16°C

Temperature in °C
—6°C— Average for January
—16°C— Average for July

Skegness

Aberystwyth

16°C

Penzance

Temp °C
25 20 15 10 5 0
Average monthly temperature
J F M A M J J A S O N D

Penzance

6°C

Warmer than 6°C in January / Warmer than 16°C in July

Colder than 6°C in January / Warmer than 16°C in July

Warmer than 6°C in January / Colder than 16°C in July

Colder than 6°C in January / Colder than 16°C in July

CLIMATE RECORDS

The climate records show some of the most extreme weather ever recorded in Britain.

Wettest Place 4391mm annual rainfall Styhead Tarn, Cumbria.

Greatest Snowfall 1524mm Upper Teesdale & Denbighshire Hills 1947

Heaviest Hailstone Horsham, W. Sussex 1958

Worst Thunderstorm Widecombe in the Moor, Devon 1638

Highest Temperature 35·9°C Cheltenham, Gloucestershire 1976

The temperature map, above, gives us information about the two things - winter temperatures and summer temperatures

Take winter temperatures first. Look for the line which shows **6° Average Temperature in January**.

This line divides Britain into two parts - places which, on average, are either warmer than 6°, or colder than 6° in January. We have to say 'on average' because it is not the same temperature all the time.

1 Using the map key; is the eastern or western part of Britain usually warmer in January?

Now take summer temperatures. The line to look for shows **16° Average Temperature in July**. The proper name for a temperature line of this sort is an **isotherm**.

2 In July is it usually warmer in northern or southern Britain?

3 Which part of Britain usually has the warmest climate all year? Which is the coldest part?

The **line graphs** show the average temperatures for each month in four places in Britain.

4 a Which place has the highest temperature in summer?

b Which place has the coldest winters?

© Collins ◇ Longman Atlases

Scale 1:2 000 000

1 centimetre on the map represents 20 kilometres on the ground.

0 20 40 60 80 km

International boundary
National boundary
County or Region boundary
Motorway
Main road
Railway
⊕ Main airport
■ City
● Large town
· Town

N O R T H   S E A

N
W — E
S

I R I S H   S E A

SCOTLAND

TAYSIDE
Tobermory
Mull
Oban
L. Awe
L. Lomond
L. Tummel
Rannoch
L. Tay
Forfar
Arbroath
Dundee
Firth of Tay
St. Andrews
Fife Ness
Crieff
Callander
Perth
Cupar
FIFE
Glenrothes
Kirkcaldy
Forth
North Berwick
CENTRAL
Stirling
Dunfermline
Firth of Forth
Grangemouth
Haddington
Falkirk
Cumbernauld
LOTHIAN
Musselburgh
Dalkeith
Livingston
Edinburgh
Dumbarton
Greenock
Paisley
Glasgow
Coatbridge
Motherwell
STRATHCLYDE
Hamilton
East Kilbride
Lanark
Peebles
Selkirk
BORDERS
Galashiels
Newtown
Boswells
Hawick
Kelso
Duns
Berwick-upon-Tweed
St. Abb's Hd.
Eyemouth
Tweed
Largs
Ardrossan
Irvine
Troon
Prestwick
Kilmarnock
Ayr
Cumnock
Sanda
Ailsa Craig
Girvan
DUMFRIES
AND
GALLOWAY
Lockerbie
Annan
Newton Stewart
Wigtown
Dee
Castle Douglas
Kirkcudbright
Dumfries
Solway Firth
Luce Bay
Mull of Galloway
Milleur Pt.
Stranraer
Campbeltown
Kintyre
Sd. of Bute
Bute
Arran
Firth of Clyde
Colonsay
Jura
Islay
Sd. of Jura

NORTHERN
IRELAND
ANTRIM
Larne
Carrickfergus
Newtown abbey
Bangor
Belfast L.
Belfast
Lisburn
Newtownards
Strangford L.
Downpatrick
DOWN
Newcastle
Portaferry

ENGLAND

NORTHUMBERLAND
Ashington
Morpeth
Tyne
Farne Islands
Tynemouth
Newcastle upon Tyne
TYNE AND WEAR
Sunderland
Washington
Peterlee
Hartlepool
CLEVELAND
Middlesbrough
Newton Aycliffe
Bishop Auckland
DURHAM
Durham
Darlington
A1(M)
Tees
Whitby
Northallerton
Swale
Ure
Wharfe
NORTH YORKSHIRE
Scarborough
Flamborough Head
Bridlington
Beverley
Kingston upon Hull
Humber
Spurn Head
Cleethorpes
Grimsby
HUMBERSIDE
Scunthorpe
M62
M18
Doncaster
M180
Trent
SOUTH YORKSHIRE
Barnsley
Rotherham
Sheffield
Worksop
Mansfield
DERBY-
Chesterfield
Matlock
Harrogate
York
Castleford
Aire
Wakefield
WEST YORKSHIRE
Leeds
Bradford
Halifax
Huddersfield
M1
LANCASHIRE
Lancaster
Morecambe Bay
Fleetwood
Blackpool
Preston
Burnley
Blackburn
Bolton
M61
M6
Wigan
Rochdale
Oldham
Salford
MANCHESTER
Manchester
St. Helens
Warrington
Stockport
Macclesfield
CHESHIRE
M56
Widnes
Runcorn
M53
Chester
Crewe
Southport
Ormskirk
Skelmersdale
Liverpool Bay
Liverpool
Birkenhead
Dee
Mold
Ruthin
Denbigh
Colwyn
Menai Str.
Anglesey
Holyhead
Bangor
Caernarfon

CUMBRIA
Carlisle
Penrith
Eden
Derwent
Windermere
Kendal
Workington
Whitehaven
Barrow-in-Furness
Ribble

ISLE OF MAN
Douglas
Calf of Man

Isle of Man

ATLANTIC OCEAN

Scale 1:2 000 000

0 20 40 60 80 km

1 centimetre on the map represents
20 kilometres on the ground

NORTH SEA

Orkney Islands
ORKNEY
N. Ronaldsay
Westray
Sanday
Rousay
Stronsay
Shapinsay
Mainland
Kirkwall
Scapa Flow
Hoy
S. Ronaldsay
Pentland Firth

SHETLAND
Shetland Islands
Same Scale
Unst
Yell
Fetlar
Whalsay
Mainland
Lerwick
Bressay
Foula
Fair Isle
Sumburgh Hd.

Flannan Is.
Butt of Lewis
Stornoway
Lewis
WESTERN
Harris
Sd. of Harris
North Uist
Lochmaddy
Benbecula
ISLES
South Uist
Lochboisdale
Eriskay
Barra
Mingulay
Canna
Uig
Portree
Skye
Kyle of Lochalsh
Rhum
Eigg
Muck
Pt. of Ardnamurchan
Coll
Tobermory
Tiree
Staffa
Iona
Mull
Colonsay
L. Linnhe
Oban
Firth of Lorn
Lochgilphead
Jura
Sd. of Jura
Islay
Loch Fyne
Loch Tyne
Sanda
Ailsa Craig
Campbeltown
Arran
Mull of Galloway

Outer Hebrides
The Minch
The Little Minch
Inner Hebrides

C. Wrath
Durness
Thurso
Dunnet Hd.
Duncansby Hd.
Wick
Lochinver
L. Shin
Lairg
Ullapool
Golspie
Dornoch Firth
Tarbat Ness
L. Maree
Invergordon
Cromarty Firth
Moray Firth
Lossiemouth
Elgin
Buckie
Banff
Fraserburgh
Dingwall
Nairn
Forres
Keith
Rattray Hd.
HIGHLAND
Inverness
Deveron
Peterhead
Buchan Ness
L. Ness
GRAMPIAN
Spey
Inverurie
Don
SCOTLAND
Kingussie
Aberdeen
Mallaig
L. Morar
Dee
Fort William
L. Rannoch
L. Tummel
N. Esk
Stonehaven
S. Esk
Montrose
L. Tay
TAYSIDE
Tay
Forfar
Arbroath
L. Awe
Dundee
Perth
Firth of Tay
Crieff
Earn
Cupar
St. Andrews
Callander
Kinross
FIFE
Fife Ness
CENTRAL
L. Lomond
Stirling
Forth
Alloa
Glenrothes
Kirkcaldy
Dunfermline
Firth of Forth
Falkirk
Grangemouth
North Berwick
Dumbarton
M80
Cumbernauld
EDINBURGH
Haddington
Dunoon
Greenock
M8
Bathgate
Livingston
Musselburgh
St. Abb's Head
Glasgow
Coatbridge
LOTHIAN
Dalkeith
Eyemouth
Rothesay
Paisley
Hamilton
Motherwell
Duns
Berwick-upon-Tweed
Largs
East Kilbride
Lanark
Tweed
Bute
Sd. of Bute
Ardrossan
Irvine
Kilmarnock
Peebles
Galashiels
Kelso
Newtown St. Boswells
Troon
Prestwick
Cumnock
Selkirk
Hawick
BORDERS
Ayr
Teviot
Farne Islands
Girvan
Nith
DUMFRIES
NORTHUMBERLAND
Ashington
Dee
AND
Dumfries
Lockerbie
ENGLAND
Milleur Pt.
GALLOWAY
Annan
Tyne
Newton Stewart
Castle Douglas
Newcastle upon Tyne
Tynemouth
Stranraer
Wigtown
Kirkcudbright
Solway Firth
Carlisle
TYNE AND WEAR
Sunderland
North Channel
Luce Bay
Eden
Washington
Durham
Bangor
Belfast Lough
Mull of Galloway
M6
Derwent
CUMBRIA
Penrith
Tees
Peterlee
Hartlepool
DURHAM
Auckland

STRATHCLYDE

N
W E
S

Legend

————	International boundary
————	National boundary
– – – –	County or region boundary
————	Motorway
————	Main road
————	Railway
⊕	Main airport
■	City
●	Large town
•	Town

Scale 1:2 000 000

0   20   40   60   80km

1 centimetre on the map represents
20 kilometres on the ground

Legend

▬▬▬	International boundary
─────	National boundary
-----	County or region boundary
═════	Motorway
─────	Main road
─────	Railway
⊕	Main airport
■	City
●	Large town
•	Town

© Collins ◊ Longman Atlases

ANCIENT AND ROMAN BRITAIN

These two photos show examples of Prehistoric and Roman sites that can still be seen today.

Ancient Britain
+ Prehistoric sites
ICENI British tribes (1st century AD)

Roman Britain
— Boundary of the Roman Empire
— Roman roads
ⅢⅢ Roman walls
● Main Roman towns
○ Ancient British towns

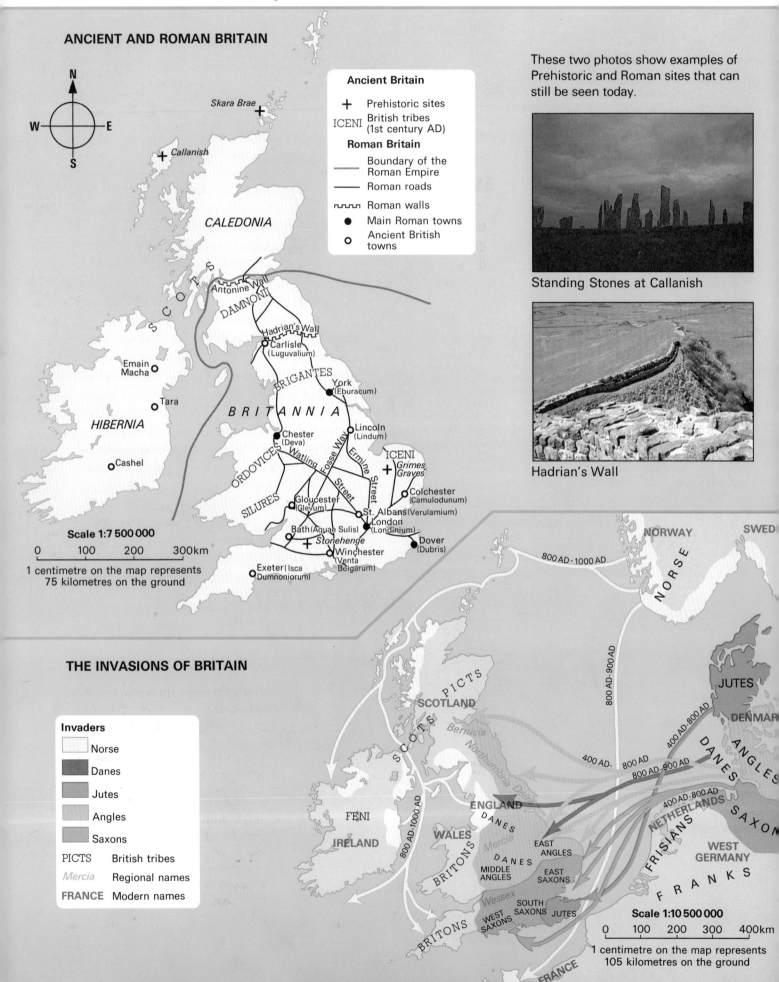

Skara Brae
Callanish
CALEDONIA

SCOTS
Antonine Wall
DAMNONI
Hadrian's Wall
Carlisle (Luguvalium)
BRIGANTES
Emain Macha
York (Eburacum)
Tara
HIBERNIA
B R I T A N N I A
Lincoln (Lindum)
Cashel
Chester (Deva)
ICENI
ORDOVICES
Watling Street
Fosse Way
Ermine Street
Grimes Graves
SILURES
Gloucester (Glevum)
Colchester (Camulodunum)
St. Albans (Verulamium)
Bath (Aquae Sulis)
London (Londinium)
Stonehenge
Dover (Dubris)
Winchester (Venta Belgarum)
Exeter (Isca Dumnoniorum)

Scale 1:7 500 000
0   100   200   300km
1 centimetre on the map represents 75 kilometres on the ground

Standing Stones at Callanish

Hadrian's Wall

THE INVASIONS OF BRITAIN

Invaders
Norse
Danes
Jutes
Angles
Saxons
PICTS British tribes
Mercia Regional names
FRANCE Modern names

NORWAY   SWEDEN
800 AD - 1000 AD
N O R S E
800 AD - 900 AD
JUTES
DENMARK
400 AD - 800 AD
D A N E S
400 AD
800 AD
800 AD - 900 AD
ANGLES
400 AD - 800 AD
PICTS
SCOTLAND
Bernicia
NETHERLANDS
SAXON
Nordhumbria Deira
FRISIANS
800 AD-1000 AD
SCOTS
ENGLAND
DANES
WEST GERMANY
FENI
WALES
Mercia
EAST ANGLES
F R A N K S
IRELAND
BRITONS
DANES
MIDDLE ANGLES
EAST SAXONS
Wessex
SOUTH SAXONS
WEST SAXONS
JUTES
BRITONS
FRANCE

Scale 1:10 500 000
0   100   200   300   400km
1 centimetre on the map represents 105 kilometres on the ground

© Collins ◇ Longman At

Medieval England and Wales

From the 12th to the 16th centuries most people in Britain lived in southeast England. Many of the industries were also found there. These industries were mainly developed from traditional crafts and used locally available raw materials such as wool.

By contrast the upland regions to the north and west were sparsely populated. However these highlands were rich in minerals. Most of the minerals mined there were exported to the south for manufacturing.

16th AND 17th CENTURY BRITAIN

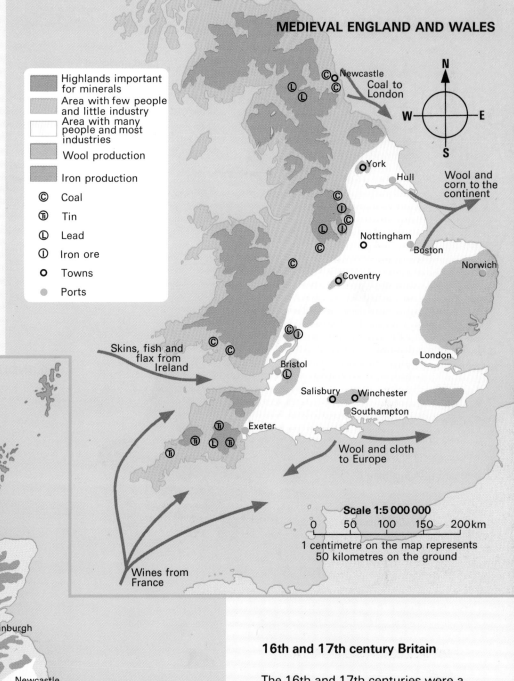

MEDIEVAL ENGLAND AND WALES

Legend:
- Highlands important for minerals
- Area with few people and little industry
- Area with many people and most industries
- Wool production
- Iron production
- © Coal
- Ⓣ Tin
- Ⓛ Lead
- Ⓘ Iron ore
- ○ Towns
- • Ports

Coal to London

Wool and corn to the continent

Skins, fish and flax from Ireland

Wool and cloth to Europe

Wines from France

Scale 1:5 000 000

0   50   100   150   200km

1 centimetre on the map represents 50 kilometres on the ground

Towns labelled: Newcastle, York, Hull, Nottingham, Boston, Norwich, Coventry, London, Bristol, Salisbury, Winchester, Southampton, Exeter

Legend for 16th and 17th century map:
- Mixed farming
- Wood pasture
- Open pasture
- Dairying
- Upland farming
- ○ Towns over 10 000 in 1700
- • Main woollen centres

Towns labelled: Glasgow, Edinburgh, Newcastle, York, Norwich, Yarmouth, Colchester, London, Bristol, Exeter

Scale 1:7 500 000

0   100   200   300km

1 centimetre on the map represents 75 kilometres on the ground

16th and 17th century Britain

The 16th and 17th centuries were a period of great change, and the population increased rapidly.

The greatest changes took place in farming. New types of farm machinery were developed, grassland was improved and new crops such as clover, turnips and potatoes were introduced. These developments along with new ideas for crop rotations, the use of fertilizers and changes in the field system all helped to increase the amount of food produced.

© Collins ◇ Longman Atlases

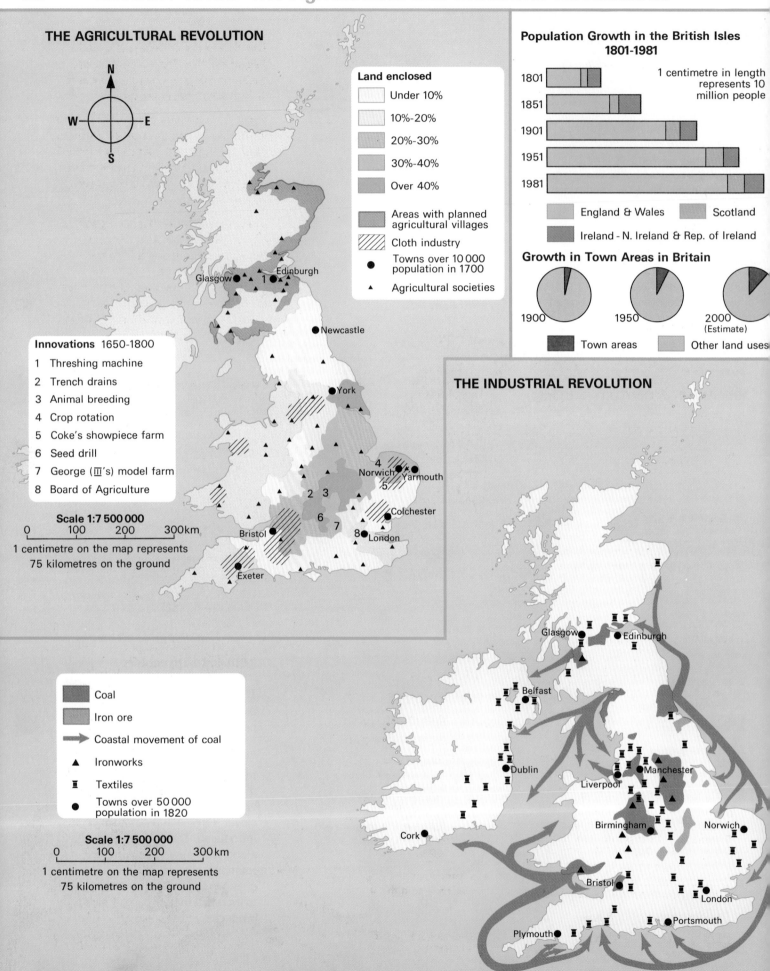

THE AGRICULTURAL REVOLUTION

N
W E
S

Land enclosed

- Under 10%
- 10%-20%
- 20%-30%
- 30%-40%
- Over 40%

Areas with planned agricultural villages

Cloth industry

● Towns over 10 000 population in 1700

▲ Agricultural societies

Innovations 1650-1800

1 Threshing machine
2 Trench drains
3 Animal breeding
4 Crop rotation
5 Coke's showpiece farm
6 Seed drill
7 George (III's) model farm
8 Board of Agriculture

Glasgow Edinburgh

Newcastle

York

Norwich Yarmouth

Colchester

Bristol

London

Exeter

Scale 1:7 500 000

0   100   200   300km

1 centimetre on the map represents
75 kilometres on the ground

Population Growth in the British Isles 1801-1981

1801
1851
1901
1951
1981

1 centimetre in length represents 10 million people

England & Wales   Scotland

Ireland - N. Ireland & Rep. of Ireland

Growth in Town Areas in Britain

1900   1950   2000 (Estimate)

Town areas   Other land uses

THE INDUSTRIAL REVOLUTION

Coal
Iron ore
→ Coastal movement of coal
▲ Ironworks
Textiles
● Towns over 50 000 population in 1820

Scale 1:7 500 000

0   100   200   300km

1 centimetre on the map represents
75 kilometres on the ground

Glasgow Edinburgh

Belfast

Dublin

Manchester
Liverpool

Birmingham   Norwich

Bristol

London

Cork

Plymouth   Portsmouth

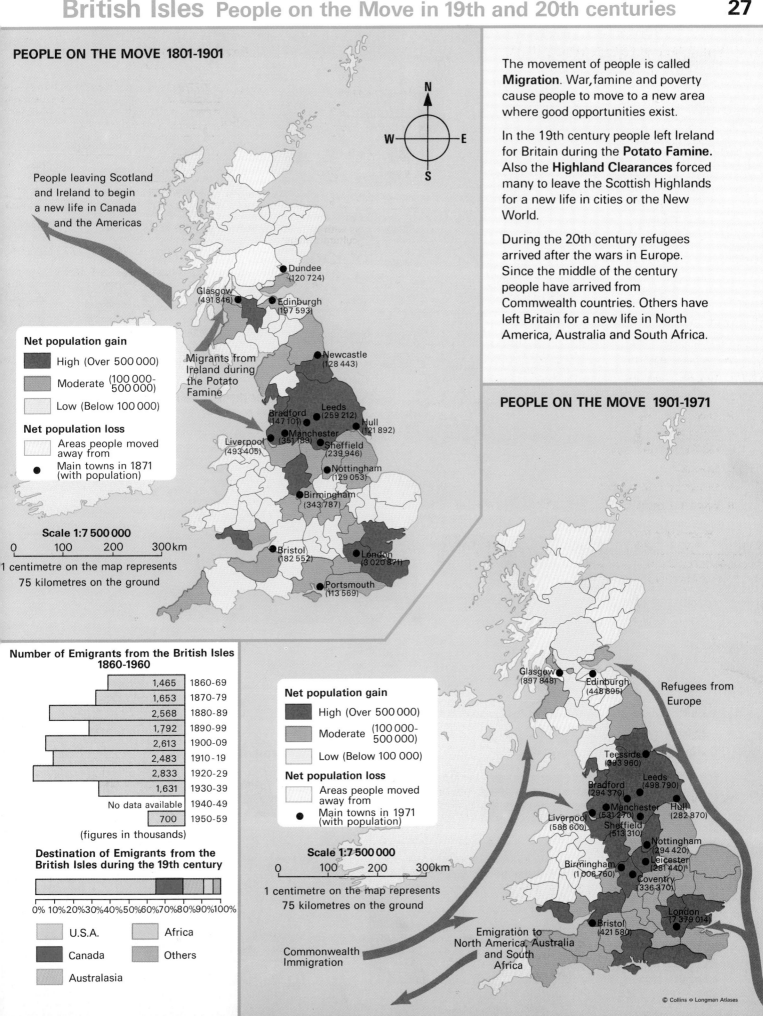

PEOPLE ON THE MOVE 1801-1901

People leaving Scotland and Ireland to begin a new life in Canada and the Americas

Migrants from Ireland during the Potato Famine

Net population gain

High (Over 500 000)	
Moderate (100 000-500 000)	
Low (Below 100 000)	

Net population loss

Areas people moved away from

● Main towns in 1871 (with population)

Scale 1:7 500 000

0   100   200   300km

1 centimetre on the map represents 75 kilometres on the ground

Dundee (120 724)
Glasgow (491 846)
Edinburgh (197 593)
Newcastle (128 443)
Leeds (259 212)
Bradford (147 101)
Hull (121 892)
Manchester (351 189)
Liverpool (493 405)
Sheffield (239 946)
Nottingham (129 053)
Birmingham (343 787)
Bristol (182 552)
London (3 020 871)
Portsmouth (113 569)

The movement of people is called **Migration**. War, famine and poverty cause people to move to a new area where good opportunities exist.

In the 19th century people left Ireland for Britain during the **Potato Famine**. Also the **Highland Clearances** forced many to leave the Scottish Highlands for a new life in cities or the New World.

During the 20th century refugees arrived after the wars in Europe. Since the middle of the century people have arrived from Commwealth countries. Others have left Britain for a new life in North America, Australia and South Africa.

PEOPLE ON THE MOVE 1901-1971

Number of Emigrants from the British Isles 1860-1960

Figures	Period
1,465	1860-69
1,653	1870-79
2,568	1880-89
1,792	1890-99
2,613	1900-09
2,483	1910-19
2,833	1920-29
1,631	1930-39
No data available	1940-49
700	1950-59

(figures in thousands)

Destination of Emigrants from the British Isles during the 19th century

0% 10% 20% 30% 40% 50% 60% 70% 80% 90% 100%

U.S.A.	Africa
Canada	Others
Australasia	

Net population gain

High (Over 500 000)	
Moderate (100 000-500 000)	
Low (Below 100 000)	

Net population loss

Areas people moved away from

● Main towns in 1971 (with population)

Scale 1:7 500 000

0   100   200   300km

1 centimetre on the map represents 75 kilometres on the ground

Refugees from Europe

Glasgow (897 848)
Edinburgh (448 895)
Teesside (393 960)
Bradford (294 370)
Leeds (498 790)
Manchester (541 270)
Hull (282 870)
Liverpool (588 600)
Sheffield (513 310)
Nottingham (294 420)
Leicester (281 440)
Birmingham (1 006 760)
Coventry (336 370)
London (7 379 014)
Bristol (421 580)

Commonwealth Immigration

Emigration to North America, Australia and South Africa

© Collins ○ Longman Atlases

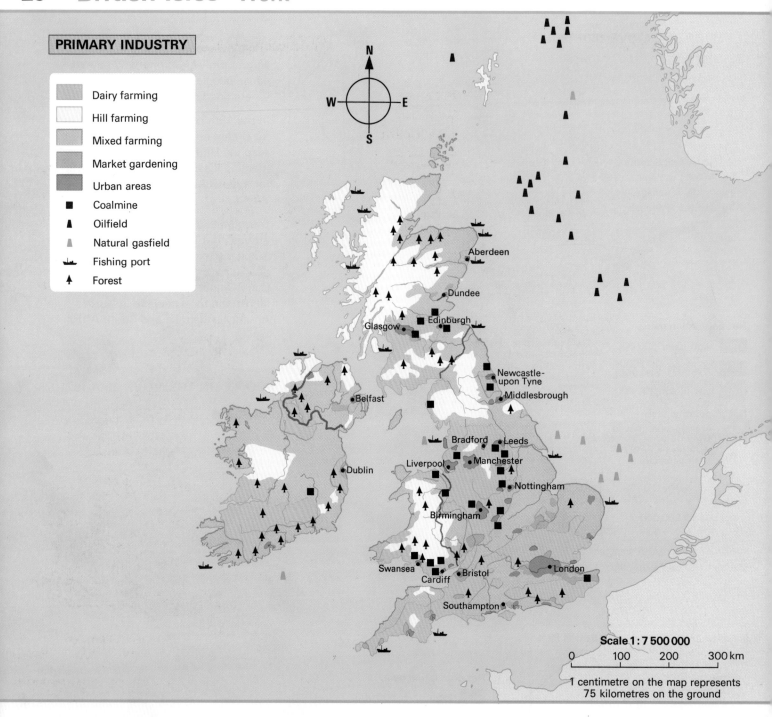

PRIMARY INDUSTRY

- Dairy farming
- Hill farming
- Mixed farming
- Market gardening
- Urban areas
- ■ Coalmine
- ▲ Oilfield
- ▲ Natural gasfield
- ⛴ Fishing port
- ♠ Forest

Aberdeen
Dundee
Edinburgh
Glasgow
Newcastle-upon Tyne
Middlesbrough
Belfast
Bradford
Leeds
Liverpool
Manchester
Dublin
Nottingham
Birmingham
Swansea
Cardiff
Bristol
London
Southampton

Scale 1 : 7 500 000

0   100   200   300 km

1 centimetre on the map represents
75 kilometres on the ground

Definitions

Primary industry is where raw materials are produced from the earth's resources, for example in forestry, farming, fishing and mining.

Secondary, or manufacturing industry, is where raw materials are turned into finished goods, for example cars, chemicals, clothing and food.

Tertiary, or service industry, is where people provide services for others, for example banking, teaching, tourism and insurance.

Employment in Britain

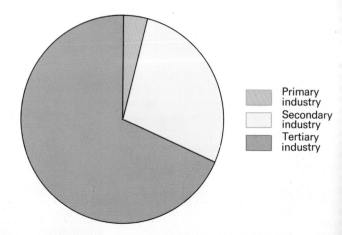

- Primary industry
- Secondary industry
- Tertiary industry

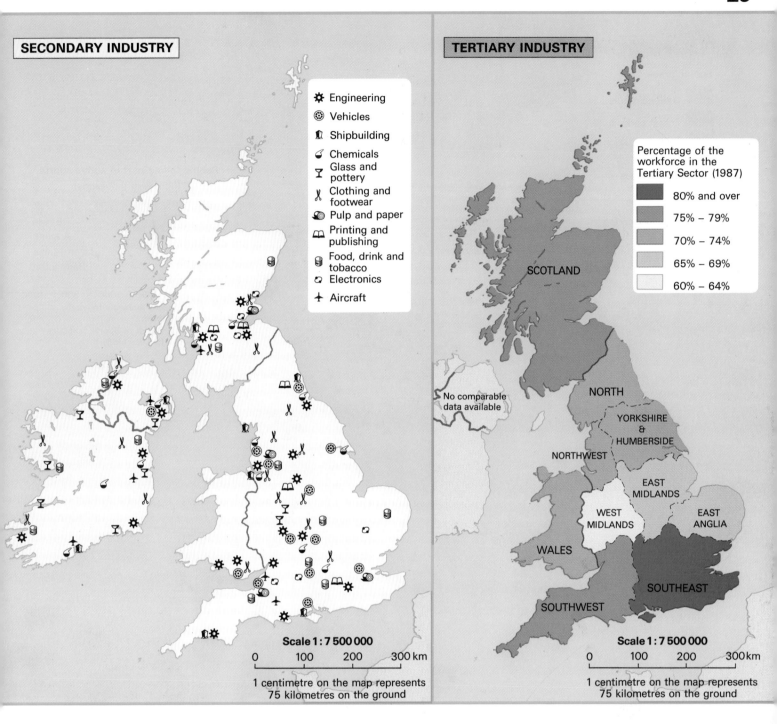

SECONDARY INDUSTRY

✳	Engineering
◉	Vehicles
⚓	Shipbuilding
◗	Chemicals
▽	Glass and pottery
✂	Clothing and footwear
◎	Pulp and paper
📖	Printing and publishing
⬚	Food, drink and tobacco
◌	Electronics
✈	Aircraft

Scale 1 : 7 500 000

0   100   200   300 km

1 centimetre on the map represents
75 kilometres on the ground

TERTIARY INDUSTRY

Percentage of the workforce in the Tertiary Sector (1987)

- 80% and over
- 75% – 79%
- 70% – 74%
- 65% – 69%
- 60% – 64%

SCOTLAND

No comparable data available

NORTH

YORKSHIRE & HUMBERSIDE

NORTHWEST

EAST MIDLANDS

WEST MIDLANDS

EAST ANGLIA

WALES

SOUTHWEST

SOUTHEAST

Scale 1 : 7 500 000

0   100   200   300 km

1 centimetre on the map represents
75 kilometres on the ground

Employment in Britain by Region

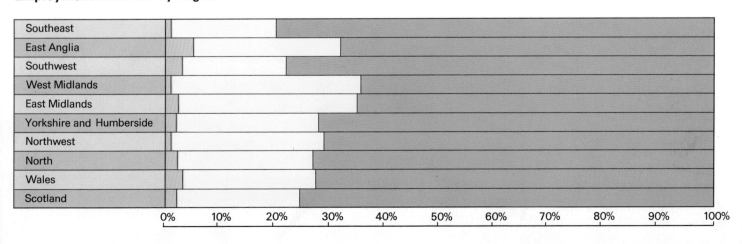

Region		
Southeast		
East Anglia		
Southwest		
West Midlands		
East Midlands		
Yorkshire and Humberside		
Northwest		
North		
Wales		
Scotland		

0%   10%   20%   30%   40%   50%   60%   70%   80%   90%   100%

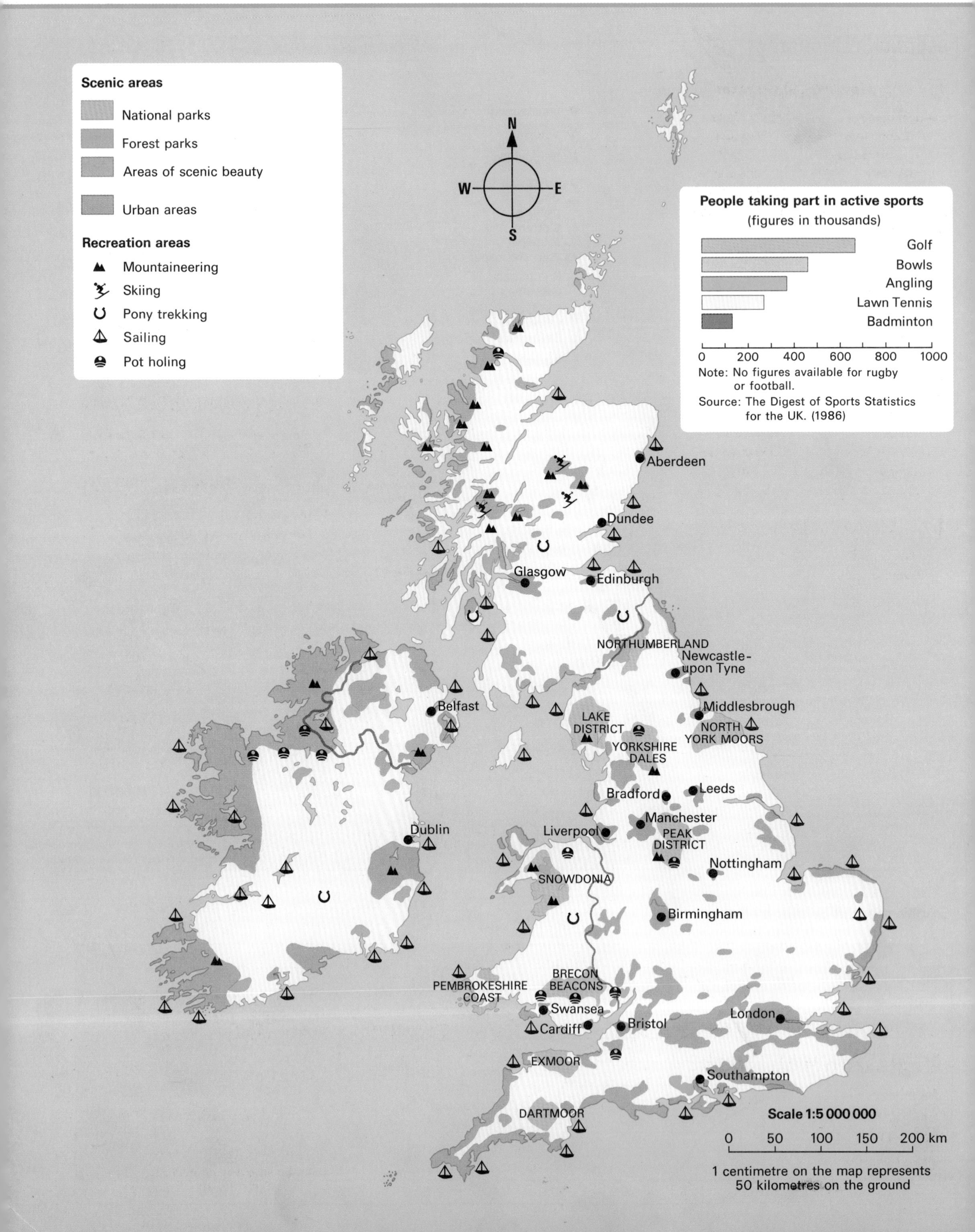

Scenic areas

National parks
Forest parks
Areas of scenic beauty
Urban areas

Recreation areas

▲▲ Mountaineering
🎿 Skiing
⛎ Pony trekking
⛵ Sailing
⬤ Pot holing

People taking part in active sports
(figures in thousands)

| Golf |
| Bowls |
| Angling |
| Lawn Tennis |
| Badminton |

0 200 400 600 800 1000

Note: No figures available for rugby or football.
Source: The Digest of Sports Statistics for the UK. (1986)

N
W — E
S

Aberdeen
Dundee
Glasgow
Edinburgh

NORTHUMBERLAND
Newcastle-upon Tyne
Middlesbrough
NORTH YORK MOORS
LAKE DISTRICT
YORKSHIRE DALES
Belfast
Leeds
Bradford
Manchester
Liverpool
PEAK DISTRICT
Nottingham
Dublin
SNOWDONIA
Birmingham
PEMBROKESHIRE COAST
BRECON BEACONS
Swansea
Cardiff
Bristol
London
EXMOOR
Southampton
DARTMOOR

Scale 1:5 000 000

0 50 100 150 200 km

1 centimetre on the map represents
50 kilometres on the ground

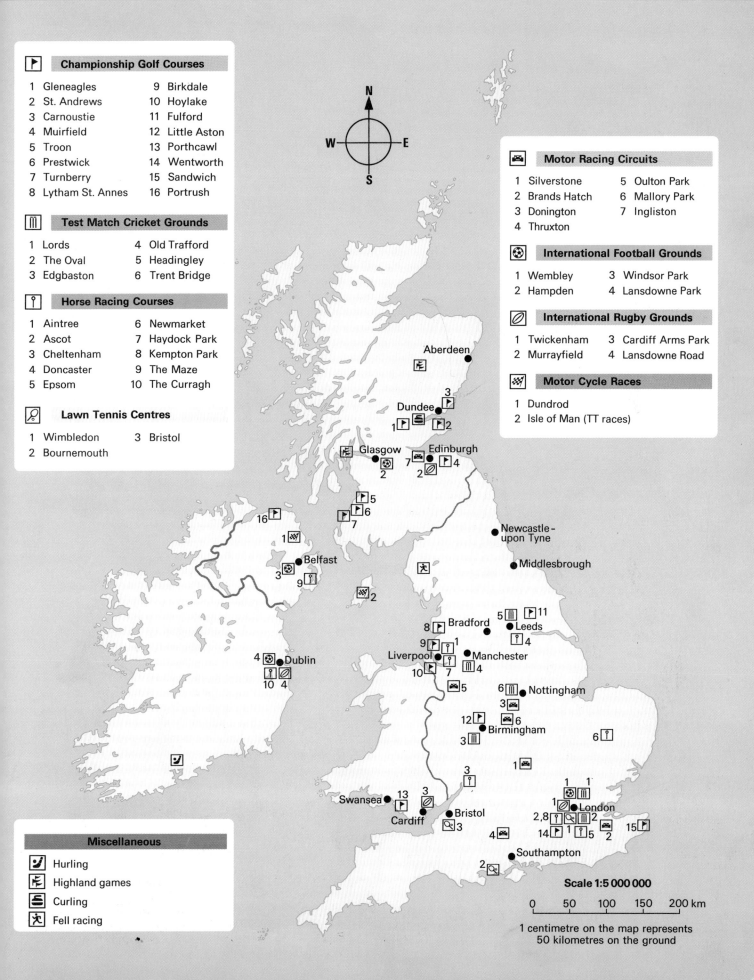

Championship Golf Courses
1 Gleneagles
2 St. Andrews
3 Carnoustie
4 Muirfield
5 Troon
6 Prestwick
7 Turnberry
8 Lytham St. Annes
9 Birkdale
10 Hoylake
11 Fulford
12 Little Aston
13 Porthcawl
14 Wentworth
15 Sandwich
16 Portrush

Test Match Cricket Grounds
1 Lords
2 The Oval
3 Edgbaston
4 Old Trafford
5 Headingley
6 Trent Bridge

Horse Racing Courses
1 Aintree
2 Ascot
3 Cheltenham
4 Doncaster
5 Epsom
6 Newmarket
7 Haydock Park
8 Kempton Park
9 The Maze
10 The Curragh

Lawn Tennis Centres
1 Wimbledon
2 Bournemouth
3 Bristol

Motor Racing Circuits
1 Silverstone
2 Brands Hatch
3 Donington
4 Thruxton
5 Oulton Park
6 Mallory Park
7 Ingliston

International Football Grounds
1 Wembley
2 Hampden
3 Windsor Park
4 Lansdowne Park

International Rugby Grounds
1 Twickenham
2 Murrayfield
3 Cardiff Arms Park
4 Lansdowne Road

Motor Cycle Races
1 Dundrod
2 Isle of Man (TT races)

Miscellaneous
Hurling
Highland games
Curling
Fell racing

Scale 1:5 000 000

0   50   100   150   200 km

1 centimetre on the map represents
50 kilometres on the ground

© Collins ◇ Longman Atlases

Ural Mountains

Kama

Pechora

Caspian Sea

Volga

ASIA

Caucasus Mts.

Novaya Zemlya

N. Dvina

Sukhona

Onega

Lake Onega

White Sea

Kola Peninsula

Volga Uplands

Don

Donets

Don

Sea of Azov

Black Sea

Central Russian Uplands

Lapland

Lake Ladoga

Dvina

Dnieper

Prut

Carpathian Mts.

Danube

Balkan Mts.

2928

Mt. Olympus

2911

Aegean Sea

Pindus Mts.

SCANDINAVIA

Gulf of Bothnia

Lake Peipus

NORTH EUROPEAN PLAIN

Bug

Dnestr

Sudeten Mts. 2664

Vistula

Hungarian Plain

Sava

Dinaric Alps

Adriatic Sea

ARCTIC OCEAN

N
E
S
W

Lofoten Is.

Vänern

Vättern

Baltic Sea

Oder

Sava

Gr. Glockner 3798

THE ALPS

Apennines

3340 Sicily

Mt. Etna

Mediterranean Sea

AFRICA

Jan Mayen

Arctic Circle

Jutland

Elbe

Rhine

Jura Mt. Blanc 4807

Corsica

Sardinia

North Sea

Seine

Loire

Massif Central 1886

Rhône

Garonne

Balearic Is.

Faroe Is.

Shetland Is.

Orkney Is.

Great Britain

Thames

English Channel

Ardennes

ICELAND Mt. Hekla 1491

BRITISH ISLES

Ireland

Bay of Biscay

Pyrenees 3404

Ebro

Cantabrian Mts. 2615

Douro

Sierra Nevada 3482

Gibraltar

MESETA

Tagus

ATLANTIC OCEAN

C. Finisterre

C. St. Vincent

Str. of Gibraltar

Legend

The colours show the height of the land above sea level.

	Over 3000 metres
	2000 - 3000 metres
	1000 - 2000 metres
	500 - 1000 metres
	200 - 500 metres
	0 - 200 metres
	Land below sea level

123 ▲   Mountain peak (height in metres)

River

Lake

Scale 1:20 000 000

0   200   400   600   800 km

1 centimetre on the map represents 200 kilometres on the ground

ARCTIC OCEAN

Svердlovsk
Perm
Ufa

Kama
Kuybyshev
Ural

Caspian Sea

UNION OF SOVIET SOCIALIST REPUBLICS

Kazan
Volga
Don
Volga

Pechora

N. Dvina
Sukhona
Onega
Gorki
Donets
Dnepropetrovsk
Don

Novaya Zemlya

White Sea
Lake Onega
Moscow
Sea of Azov

Lake Ladoga
Kharkov
Kiev
Dnieper
Odessa
Black Sea

ASIA

Leningrad
Minsk
Dvina
Lake Peipus

Istanbul
TURKEY

FINLAND
Helsinki

ROMANIA
Bucharest
Sofia
BULGARIA

Aegean Sea
Rhodes
Crete

Prut
Danube
Bug

Athens
GREECE

NORWAY
SWEDEN
Gulf of Bothnia

Stockholm
Warsaw
Vistula
POLAND
Oder
CZECHOSLOVAKIA
Prague
Budapest
HUNGARY
Belgrade
YUGOSLAVIA
ALBANIA
Tiranë

Sava
Danube
Adriatic Sea

Vänern
Vättern

Baltic Sea

Oslo

EAST GERMANY
Berlin
Elbe
Vienna
AUSTRIA
SAN MARINO

ITALY

DENMARK
Copenhagen
Hamburg
WEST GERMANY
Bonn
Rhine
Munich
Danube
Berne
SWITZ.
Milan
Po
Turin
MONACO
Rome
Naples
Palermo
Sicily
MALTA

Lyon

Jan Mayen (Nor.)

Faroe Is. (Den.)
Shetland Is.
Orkney Is.

North Sea
Edinburgh
UNITED KINGDOM
Amsterdam
NETH.
Rotterdam
BEL.
Brussels
LUX.
Paris
Seine
FRANCE
Loire
Rhône
Marseille
Corsica (Fr.)
Sardinia (It.)
Cagliari

Mediterranean Sea

Lofoten Is.
Arctic Circle

Belfast
Dublin
REPUBLIC OF IRELAND
Thames
London
English Channel
Channel Is. (U.K.)
Bay of Biscay
Garonne

ANDORRA
Barcelona
Balearic Is. (Sp.)
Palma

ICELAND
Reykjavik

Ebro
Madrid
SPAIN
Douro
Tagus
Lisbon
PORTUGAL
Gibraltar (U.K.)
Gibraltar
Str. of Gibraltar

ATLANTIC OCEAN

AFRICA

Cities and towns
■ Capital cities
● Important towns

BEL. : BELGIUM
L. : LIECHTENSTEIN
LUX. : LUXEMBOURG
NETH. : NETHERLANDS
SWITZ. : SWITZERLAND

Scale 1:20 000 000
0 200 400 600 800 km

1 centimetre on the map represents
200 kilometres on the ground

© Collins ○ Longman Atlases

East of Greenwich 0° Greenwich West of Greenwich

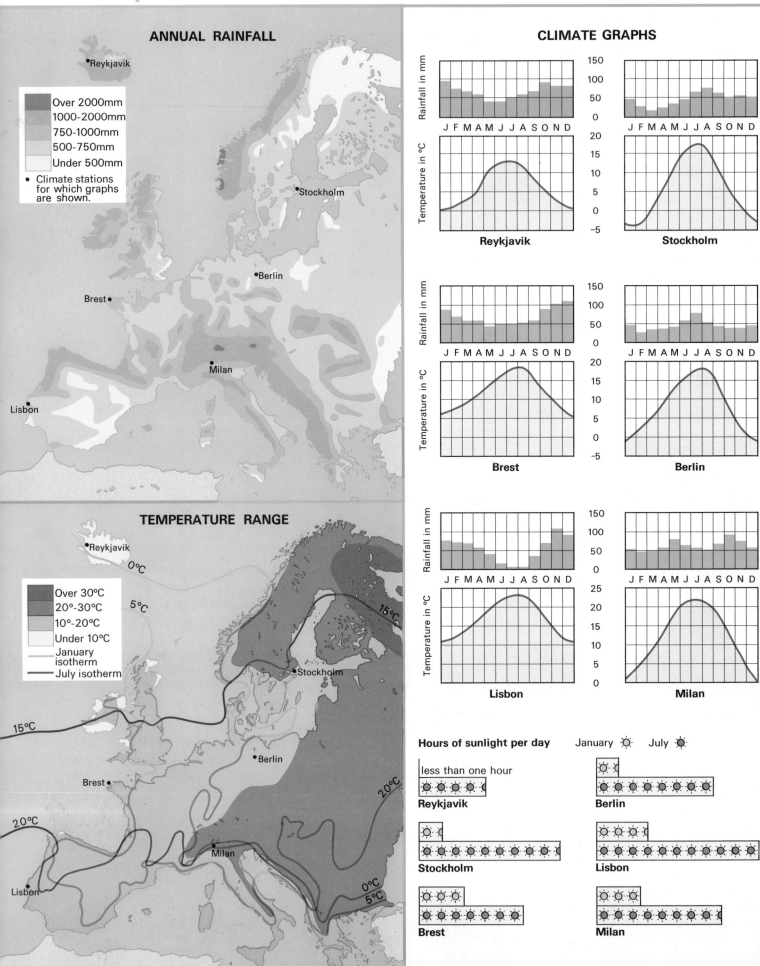

ANNUAL RAINFALL

Over 2000mm
1000-2000mm
750-1000mm
500-750mm
Under 500mm
• Climate stations for which graphs are shown.

• Reykjavik
• Stockholm
• Berlin
Brest •
Milan
Lisbon

TEMPERATURE RANGE

• Reykjavik   0°C
Over 30°C
20°-30°C
10°-20°C
Under 10°C
January isotherm
July isotherm

5°C
15°C
Stockholm
15°C
• Berlin
Brest •
20°C
20°C
Milan
Lisbon
0°C
5°C

© Collins ◊ Longman Atlases

CLIMATE GRAPHS

Rainfall in mm — 150, 100, 50, 0
J F M A M J J A S O N D

Temperature in °C — 20, 15, 10, 5, 0, -5

Reykjavik

Stockholm

Brest

Berlin

Lisbon

Milan

Hours of sunlight per day   January ☼   July ☀

less than one hour

Reykjavik

Berlin

Stockholm

Lisbon

Brest

Milan

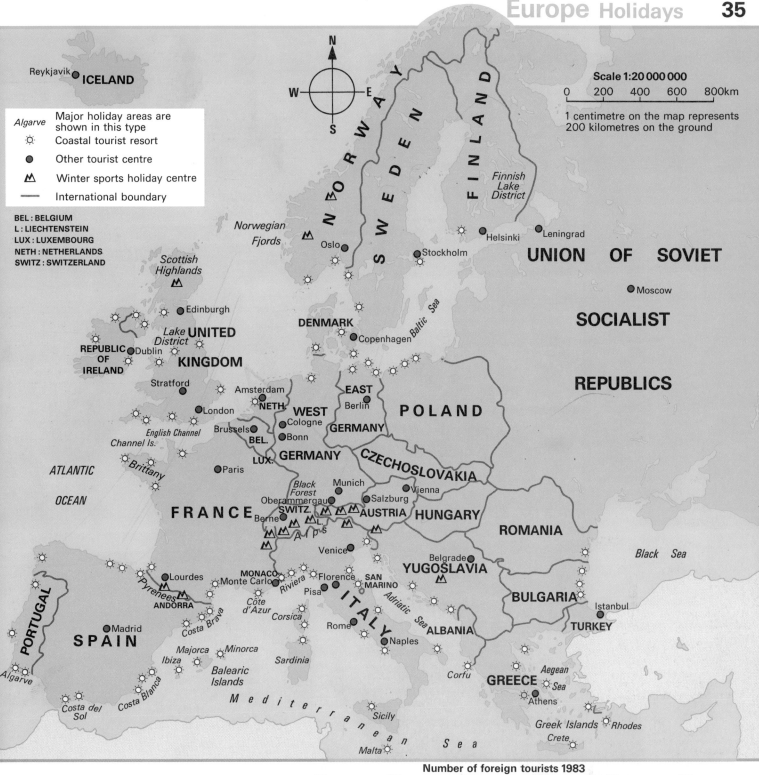

Scale 1:20 000 000

0   200   400   600   800km

1 centimetre on the map represents
200 kilometres on the ground

Algarve Major holiday areas are
shown in this type
☼ Coastal tourist resort
● Other tourist centre
⛰ Winter sports holiday centre
— International boundary

BEL : BELGIUM
L : LIECHTENSTEIN
LUX : LUXEMBOURG
NETH : NETHERLANDS
SWITZ : SWITZERLAND

Europe's main tourist areas are in the
Mediterranean countries where hot
sun and blue skies are normal in the
main holiday months of July and
August. In winter they also have
skiing centres with warmer conditions
than those further north. Figures for
foreign tourists visiting Europe in
1983 shown opposite exclude arrivals
of nationals living abroad.

Number of foreign tourists 1983

	millions of tourists per year					
more than 20M	Austria 14·5M	Belgium 6·6M	Bulgaria 5·8M	Czecho-slovakia 4·6M	Denmark 3·8M	Finland 451 000
10–20M	France 33·6M	Greece 4·8M	Hungary 6·8M	Iceland 77 600	Ireland 2·3M	Italy 22·1M
1–10M	Malta 490 800	Netherlands 3·0M	Norway 474 900	Poland 1·9M	Portugal 3·7M	Romania 5·8M
less than 1M	Spain 25·6M	Sweden 3·4M	Switzerland 9·2M	United Kingdom 12·5M	West Germany 11·3M	Yugoslavia 5·9M

© Collins ◇ Longman Atlases

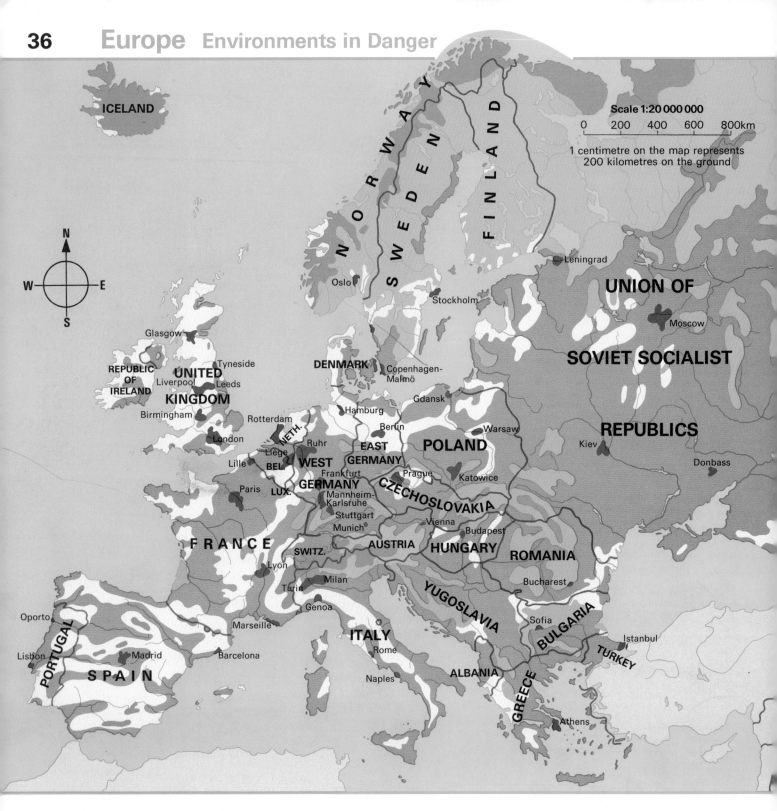

Scale 1:20 000 000

0   200   400   600   800km

1 centimetre on the map represents
200 kilometres on the ground

Environment is now a common word. What exactly does it mean?
The environment is everything that surrounds you – buildings, open spaces, grass and plants, and the air.

Some environments, like mountains, deserts and oceans have been very little affected by people, they are called **natural environments**. Other environments, such as towns and farmland, have been changed by people to suit their needs.

Mostly natural environments

Coniferous forest. Pine, spruce and other cone-bearing trees grow in this area which has a cool, wet climate.

Mixed forest. Cool, moist areas which are covered with forests of evergreen and deciduous trees.

Mountain and tundra. There are no trees growing here. Mountain areas are too high and tundra is frozen for much of the year.

Environments much affected by people

Arable land. This is farmland which is used for growing crops. It is usually more fertile than grazing land.

Grazing land. Most of this land is used as pasture and for rearing animals. Not all parts are equally fertile.

Urban area. This is an environment almost completely made by people. It includes buildings, parks and roads.

The smoke from industry can pollute the
environment of the air, or atmosphere. **Acid rain** is
a type of air pollution. When industries burn coal or
oil they let a gas called sulphur dioxide escape from
their factory chimneys. Sulphur dioxide combines
with water in the air to form an acid chemical.

This acid in the atmosphere can be carried long
distances by the wind before being washed down to
the ground as 'acid rain' or snow. This means that
countries which do not produce the chemicals can
still be affected by acid rain. Acid rain is very
harmful to the natural environment, because it kills
trees and fish in lakes and rivers.

How acid is rainwater

High

Medium   Low

Direction of wind

Lakes and streams with much
damage caused by acid rain.

Forest areas with a lot of
damage caused by acid rain.

Country that receives more than one million
tonnes of sulphur dioxide in a year.

Country that gives out more than one million
tonnes of sulphur dioxide in a year

ACID RAIN

There are many other types of air and land pollution.
• When large amounts of chemical fertilizers are
used in place of natural fertilizers, the land and
rivers can be polluted by the chemical fertilizer that
is not used up by plants.
• The sea and rivers can be polluted when
industries use them to dump unwanted waste
materials.

Land pollution
(amounts of chemical fertilisers used)

Very high   Medium

High   Low

Sea pollution

High

Medium   Low

Heavily polluted river

Oil spillage from ship

Industrial chemical accident

Nuclear power station accident

Major oil rig blowout

OTHER POLLUTION

© Collins ◊ Longman Atlases

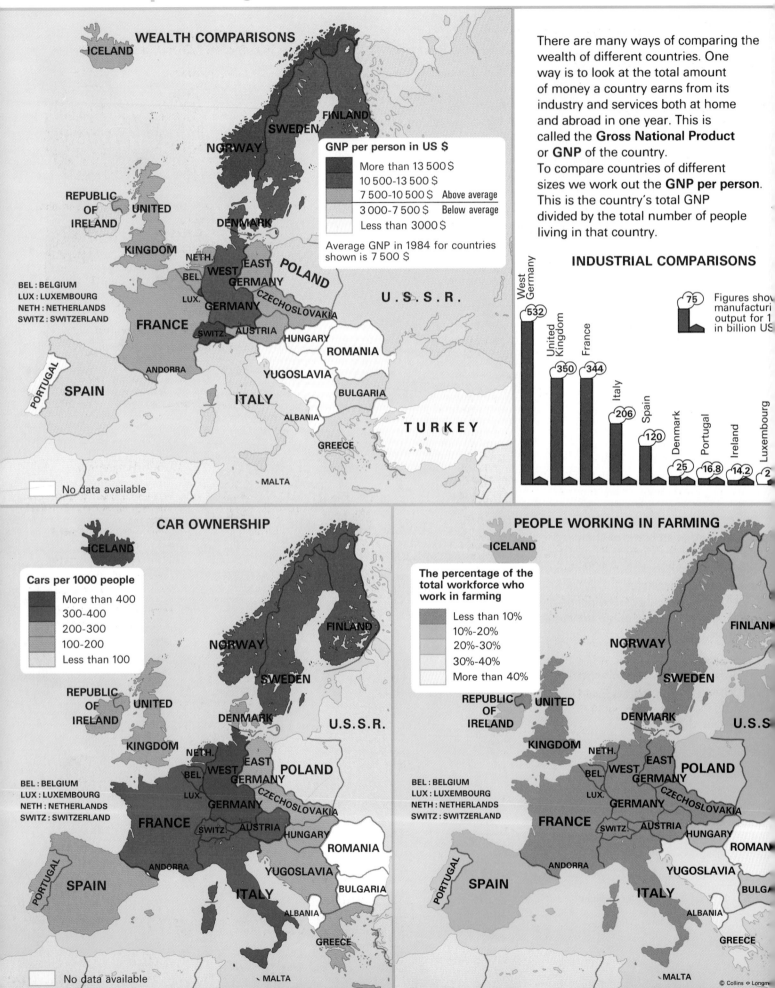

WEALTH COMPARISONS

ICELAND

FINLAND
SWEDEN
NORWAY

REPUBLIC OF IRELAND
UNITED KINGDOM
DENMARK
NETH.
EAST
BEL.
WEST GERMANY
POLAND
LUX.
GERMANY
CZECHOSLOVAKIA
U.S.S.R.
FRANCE
SWITZ. AUSTRIA
HUNGARY
ROMANIA
ANDORRA
YUGOSLAVIA
BULGARIA
PORTUGAL
SPAIN
ITALY
ALBANIA
TURKEY
GREECE
MALTA

GNP per person in US $

More than 13 500 $	
10 500-13 500 $	
7 500-10 500 $	Above average
3 000-7 500 $	Below average
Less than 3000 $	

Average GNP in 1984 for countries shown is 7 500 $

BEL : BELGIUM
LUX : LUXEMBOURG
NETH : NETHERLANDS
SWITZ : SWITZERLAND

No data available

There are many ways of comparing the wealth of different countries. One way is to look at the total amount of money a country earns from its industry and services both at home and abroad in one year. This is called the **Gross National Product** or **GNP** of the country.

To compare countries of different sizes we work out the **GNP per person**. This is the country's total GNP divided by the total number of people living in that country.

INDUSTRIAL COMPARISONS

West Germany 532
United Kingdom 350
France 344
Italy 206
Spain 120
Denmark 25
Portugal 16.8
Ireland 14.2
Luxembourg 2

75 Figures show manufacturi output for 1 in billion US

CAR OWNERSHIP

ICELAND

Cars per 1000 people

More than 400	
300-400	
200-300	
100-200	
Less than 100	

NORWAY
FINLAND
SWEDEN
DENMARK
REPUBLIC OF IRELAND
UNITED KINGDOM
U.S.S.R.
NETH.
EAST
BEL.
WEST GERMANY
POLAND
LUX.
GERMANY
CZECHOSLOVAKIA
FRANCE
SWITZ. AUSTRIA
HUNGARY
ROMANIA
ANDORRA
YUGOSLAVIA
SPAIN
ITALY
BULGARIA
ALBANIA
GREECE
MALTA

BEL : BELGIUM
LUX : LUXEMBOURG
NETH : NETHERLANDS
SWITZ : SWITZERLAND

No data available

PEOPLE WORKING IN FARMING

ICELAND

The percentage of the total workforce who work in farming

Less than 10%	
10%-20%	
20%-30%	
30%-40%	
More than 40%	

NORWAY
FINLAN
SWEDEN
DENMARK
U.S.S
REPUBLIC OF IRELAND
UNITED KINGDOM
NETH.
EAST
BEL.
WEST GERMANY
POLAND
LUX.
GERMANY
CZECHOSLOVAKIA
FRANCE
SWITZ. AUSTRIA
HUNGARY
ROMAN
ANDORRA
YUGOSLAVIA
SPAIN
ITALY
BULGA
ALBANIA
GREECE
MALTA

BEL : BELGIUM
LUX : LUXEMBOURG
NETH : NETHERLANDS
SWITZ : SWITZERLAND

© Collins ◇ Longm

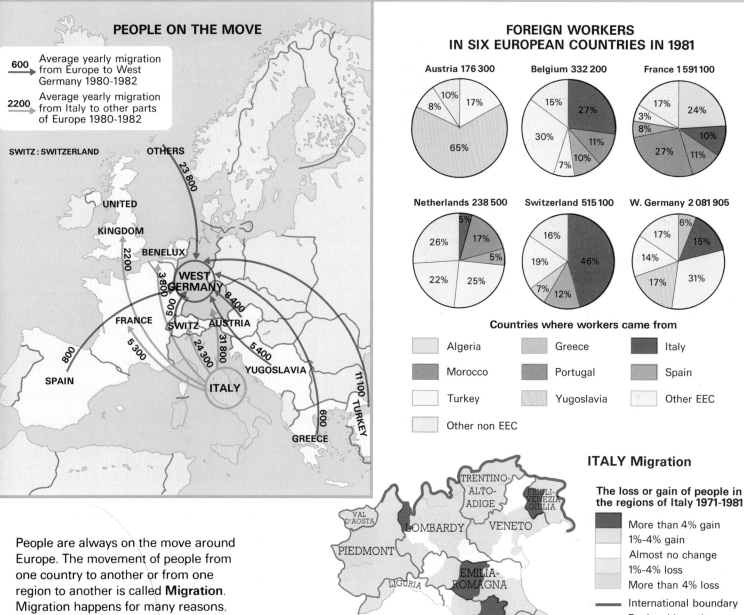

PEOPLE ON THE MOVE

600 → Average yearly migration from Europe to West Germany 1980-1982

2200 → Average yearly migration from Italy to other parts of Europe 1980-1982

SWITZ : SWITZERLAND

OTHERS 23 800

UNITED KINGDOM

BENELUX 2200

WEST GERMANY 3 800   500

FRANCE

SWITZ   AUSTRIA   8 400

5 300   24 300   31 800   5 400

SPAIN 800

ITALY

YUGOSLAVIA

11 100 TURKEY

GREECE 600

FOREIGN WORKERS IN SIX EUROPEAN COUNTRIES IN 1981

Austria 176 300
- 17%
- 10%
- 8%
- 65%

Belgium 332 200
- 15%
- 27%
- 30%
- 7%
- 10%
- 11%

France 1 591 100
- 17%
- 24%
- 3%
- 8%
- 10%
- 27%
- 11%

Netherlands 238 500
- 5%
- 26%
- 17%
- 5%
- 22%
- 25%

Switzerland 515 100
- 16%
- 19%
- 46%
- 7%
- 12%

W. Germany 2 081 905
- 6%
- 17%
- 15%
- 14%
- 17%
- 31%

Countries where workers came from
- Algeria
- Greece
- Italy
- Morocco
- Portugal
- Spain
- Turkey
- Yugoslavia
- Other EEC
- Other non EEC

People are always on the move around Europe. The movement of people from one country to another or from one region to another is called **Migration**. Migration happens for many reasons.

People move from one country to work in another country. Often this is a temporary migration as after a time they move back to where they came from. These people are sometimes called Guestworkers.

Some migrations are permanent. People decide to go and live in another country or another region within a country. This form of migration is happening in Italy for example.

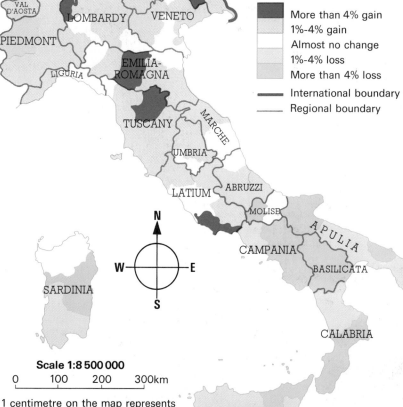

ITALY Migration

The loss or gain of people in the regions of Italy 1971-1981

- More than 4% gain
- 1%-4% gain
- Almost no change
- 1%-4% loss
- More than 4% loss
- International boundary
- Regional boundary

VAL D'AOSTA

PIEDMONT

LOMBARDY

TRENTINO-ALTO-ADIGE

VENETO

FRIULI-VENEZIA GIULIA

LIGURIA

EMILIA-ROMAGNA

TUSCANY

MARCHE

UMBRIA

LATIUM

ABRUZZI

MOLISE

CAMPANIA

APULIA

BASILICATA

SARDINIA

CALABRIA

SICILY

N W–E S

Scale 1:8 500 000

0   100   200   300km

1 centimetre on the map represents 85 kilometres on the ground

© Collins ◊ Longman Atlases

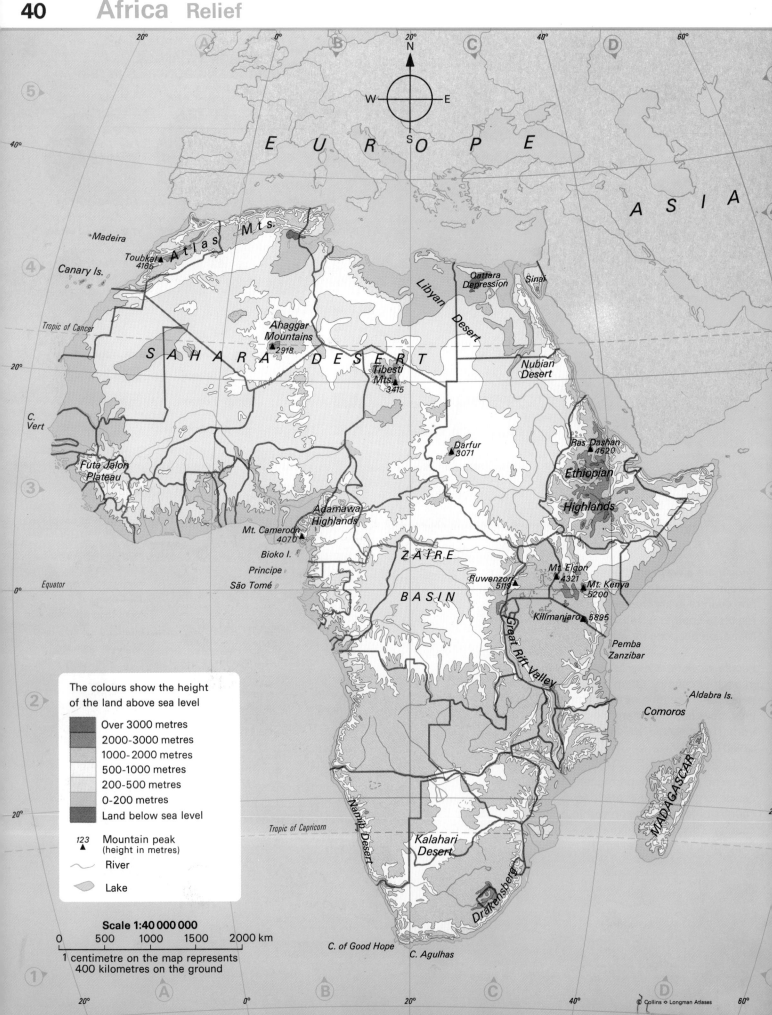

N
W E
S

EUROPE

ASIA

Madeira
Toubkal 4165
Canary Is.
Atlas Mts.
Tropic of Cancer
C. Vert
SAHARA DESERT
Ahaggar Mountains 2918
Libyan Desert
Qattara Depression
Sinai
Nubian Desert
Tibesti Mts. 3415
20°
Futa Jalon Plateau
Darfur 3071
Ras Dashan 4620
Ethiopian Highlands
Adamawa Highlands
Mt. Cameroon 4070
Bioko I.
Principe
São Tomé
Equator
ZAIRE BASIN
Mt Elgon 4321
Ruwenzori 5119
Mt. Kenya 5200
Kilimanjaro 5895
Pemba
Zanzibar
Great Rift Valley
Aldabra Is.
Comoros
MADAGASCAR
Namib Desert
Kalahari Desert
Tropic of Capricorn
Drakensberg
C. of Good Hope
C. Agulhas

The colours show the height
of the land above sea level

	Over 3000 metres
	2000-3000 metres
	1000-2000 metres
	500-1000 metres
	200-500 metres
	0-200 metres
	Land below sea level

123 ▲ Mountain peak
(height in metres)

River

Lake

Scale 1:40 000 000

0   500   1000   1500   2000 km

1 centimetre on the map represents
400 kilometres on the ground

© Collins ○ Longman Atlases

Africa Countries

E U R O P E

ASIA

Mediterranean Sea

Madeira (Port.)

Algiers • Tunis
Rabat • TUNISIA
Casablanca • Tripoli
MOROCCO Benghazi
Alexandria
Cairo
El Gîza
Canary Is. (Sp.)
El Aaiún
WESTERN SAHARA
ALGERIA
LIBYA
EGYPT
Nile
Red Sea
Tropic of Cancer

MAURITANIA
Nouakchott
MALI NIGER
Dakar Sénégal
SENEGAL Bamako Niamey CHAD
Banjul GAMBIA Ouagadougou L. Chad Khartoum
GUINEA BISSAU Bissau BURKINA SUDAN DJIBOUTI
GUINEA Niger N'Djamena Djibouti
Conakry BENIN
SIERRA LEONE Freetown IVORY COAST GHANA TOGO NIGERIA Abuja Addis Ababa
Yamoussoukro Ibadan CENTRAL AFRICAN ETHIOPIA SOMALI REPUBLIC
Monrovia Lagos REPUBLIC
LIBERIA Abidjan Accra CAMEROON Bangui
Lomé Porto Novo Yaoundé Mogadishu
Gulf of Guinea SÃO TOMÉ & PRINCIPE EQUATORIAL GUINEA L. Turkana Equator
Libreville GABON CONGO UGANDA KENYA
ZAÏRE Zaire Kampala L. Victoria Nairobi
RWANDA Mombasa
Brazzaville BURUNDI
Kinshasa L. Tanganyika Dodoma INDIAN OCEAN
ATLANTIC Kananga TANZANIA Dar es Salaam
OCEAN Luanda Aldabra Is. (Sey.)
• Ascension (U.K.)
ANGOLA MALAWI L. Malawi COMOROS
Lilongwe
ZAMBIA
Lusaka M O Z A M B I Q U E
• St. Helena (U.K.) Zambezi MADAGASCAR
Harare
ZIMBABWE Beira
NAMIBIA BOTSWANA Mozambique Channel
Walvis Bay Windhoek Antananarivo
Gaborone Tropic of Capricorn
Pretoria
Johannesburg Maputo
Orange Mbabane SWAZILAND
REPUBLIC OF Maseru Durban
SOUTH AFRICA LESOTHO
Cape Town

Cities and towns

■ Capital cities

● Important towns

Scale 1:40 000 000

0   500   1000   1500   2000 km

1 centimetre on the map represents
400 kilometres on the ground

Until the beginning of this century Europeans thought of Africa as the 'Dark Continent'. This was not because most of the people who live there are black, but because the Europeans had not managed to explore it fully. Africa was not easy to explore. In the north was the vast, almost waterless Sahara, the world's largest desert. Around the equator lay huge areas of dense jungle - the tropical rainforest. Between desert and jungle stretched the savanna grasslands, with their huge herds of wild animals.

Today Africa is fully mapped, and can be photographed in detail from orbiting satellites.

The map, on the opposite page, shows the sort of vegetation that grows in different parts of Africa. In the north, coloured orange, is the desert. Although this is very dry, a few plants can survive here. Near the equator, coloured dark green, is an area with a very different sort of vegetation. Here the climate is much wetter, and thousands of types of trees and plants grow to form dense tropical rainforest.

1 Study the vegetation map carefully:

   a What sort of vegetation is usually found around the edges of the desert?

   b What sort of vegetation surrounds the rainforest?

2 Why is the vegetation in the areas shaded red different from what is found in surrounding areas?

SATELLITE PHOTO ▶

This satellite photo was taken from 'Meteosat'. On the photo you can also see some of the changes in vegetation, although the real colours are not quite as clear as the colours on the map.

Look at the central part of Africa. Pick out the darker green colours of the rainforest, you can just see it below the cloud. Notice how this gradually changes to the greeny-brown and brown of the savanna grasslands. Then notice how this in turn gradually changes to the lighter sandy colours of the desert.

> **What the photo tells us is that the type of vegetation changes gradually. There is not that sharp change that a map seems to show.**

3 Which other continents can you see in the satellite photo?
  From the colours on the photo, describe the type of vegetation you think they have?

ATLAS SKILLS

4 a Use the map of Africa on page 40 to find the location of these lakes and rivers:

   Lake Victoria   River Cubango

   Lake Kariba   River Nile

   Lake Volta   River Zaire

b Which of these features can you pick out on the satellite photo on this page?

c In the photo there is an island off the south east coast of Africa, what is the name of the island?

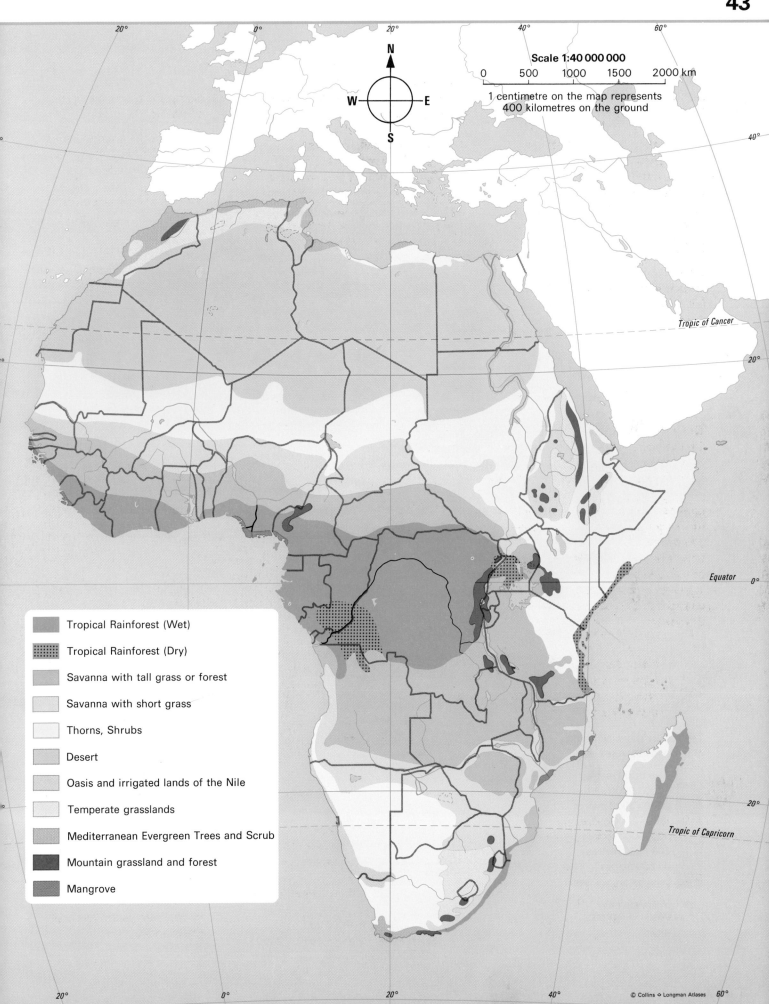

Scale 1:40 000 000

0   500   1000   1500   2000 km

1 centimetre on the map represents
400 kilometres on the ground

Tropic of Cancer

Equator

Tropic of Capricorn

Tropical Rainforest (Wet)

Tropical Rainforest (Dry)

Savanna with tall grass or forest

Savanna with short grass

Thorns, Shrubs

Desert

Oasis and irrigated lands of the Nile

Temperate grasslands

Mediterranean Evergreen Trees and Scrub

Mountain grassland and forest

Mangrove

© Collins ◇ Longman Atlases

The colours show the height
of the land above sea level

Over 5000 metres
3000 - 5000 metres
2000 - 3000 metres
1000 - 2000 metres
500 - 1000 metres
200 - 500 metres
0 - 200 metres
Land below sea level

123 Mountain peak
▲ (height in metres)

River

Lake

Scale 1:45 000 000

0   500   1000   1500   2000 km

1 centimetre on the map represents
450 kilometres on the ground

© Collins ◇ Longman Atlases

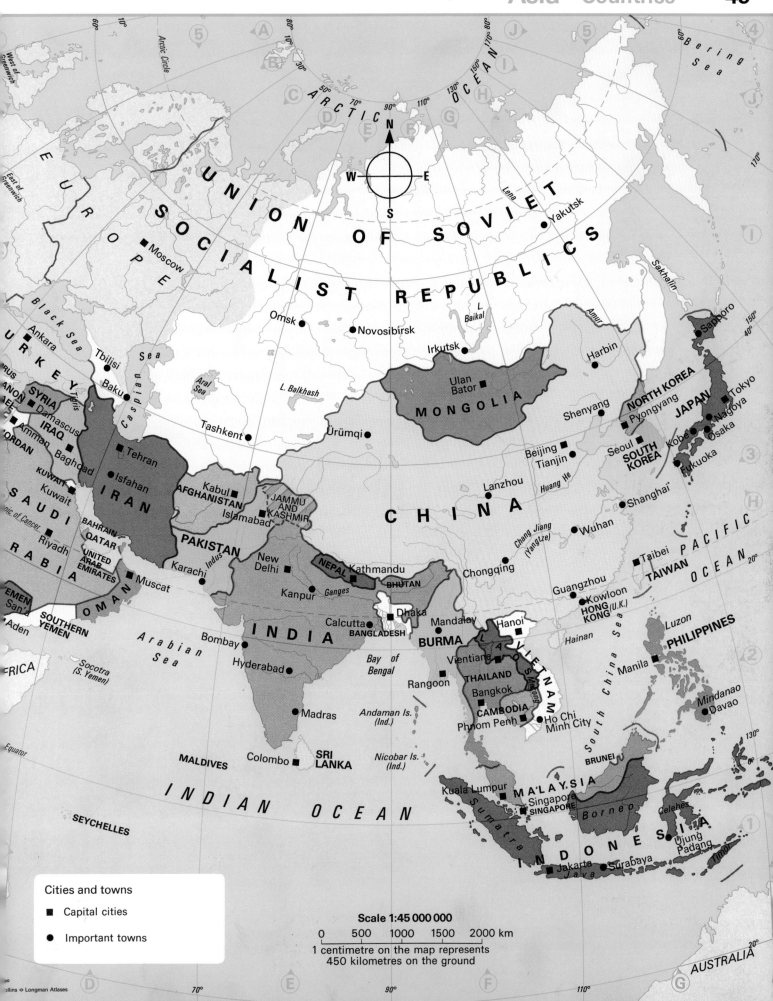

Cities and towns

■ Capital cities

● Important towns

Scale 1:45 000 000

0   500   1000   1500   2000 km

1 centimetre on the map represents
450 kilometres on the ground

Collins © Longman Atlases

G
Equator

N · E · S · W

5

4

Tropic of Capricorn

3

2

East Cape

Chatham I.

1

F
Santa Cruz Is.

New Hebrides

Loyalty Is.

New Caledonia

P A C I F I C
O C E A N

North Cape

Cook Str.

North Island

NEW
ZEALAND

South Island

E
New Ireland

New Britain

Solomon Islands

Bougainville

Guadalcanal

Coral Sea

Tasman Sea

Mt. Cook Alps
3764

Southern Alps

Stewart I.

D
Admiralty Is.

Bismarck Sea

Owen Stanley Range

Gulf of Papua

C. York

Great Barrier Reef

Mitchell

Flinders

Great Dividing Range

Darling Downs

Warrego

Blue Mts.

Botany Bay

Mt. Kosciusko
2230

C. Howe

Bass Str. Flinders I.

South East Cape

Mt. Ossa
1617
Tasmania

Scale 1:27 000 000

| 0 | 400 | 800 | 1200 km |

1 centimetre on the map represents
270 kilometres on the ground

C
Sepik

New Guinea

Fly

Torres Str.

Gulf of Carpentaria

Great Artesian Basin

Grey Range

Cooper Creek

L. Eyre

Flinders Range

Darling

Murrumbidgee

Murray

Spencer Gulf

Kangaroo I.

Puntjak Jaya
5030

Arafura Sea

Melville Is.

Arnhem Land

Barkly Tableland

Georgina

Macdonnell Ranges

Mt. Woodroffe
1515

Musgrave Ranges

Great Victoria Desert

Nullarbor Plain

L. Torrens

L. Gairdner

Great Australian Bight

B
Ceram

Timor Sea

King Leopold Ranges

Great Sandy Desert

L. Mackay

L. Disappointment

Gibson Desert

Mt. Bruce
1227

Hamersley Range

Ashburton

Darling Range

C. Leeuwin

North West Cape

I N D I A N   O C E A N

The colours show the height
of the land above sealevel.

Over 3000 metres
2000-3000 metres
1000-2000 metres
500-1000 metres
200-500 metres
0-200 metres
Land below sea level

123 ▲ Mountain peak
(height in metres)

River

Lake

© Collins ◇ Longman Atlases

A
A S I A

5

4

3

2

1

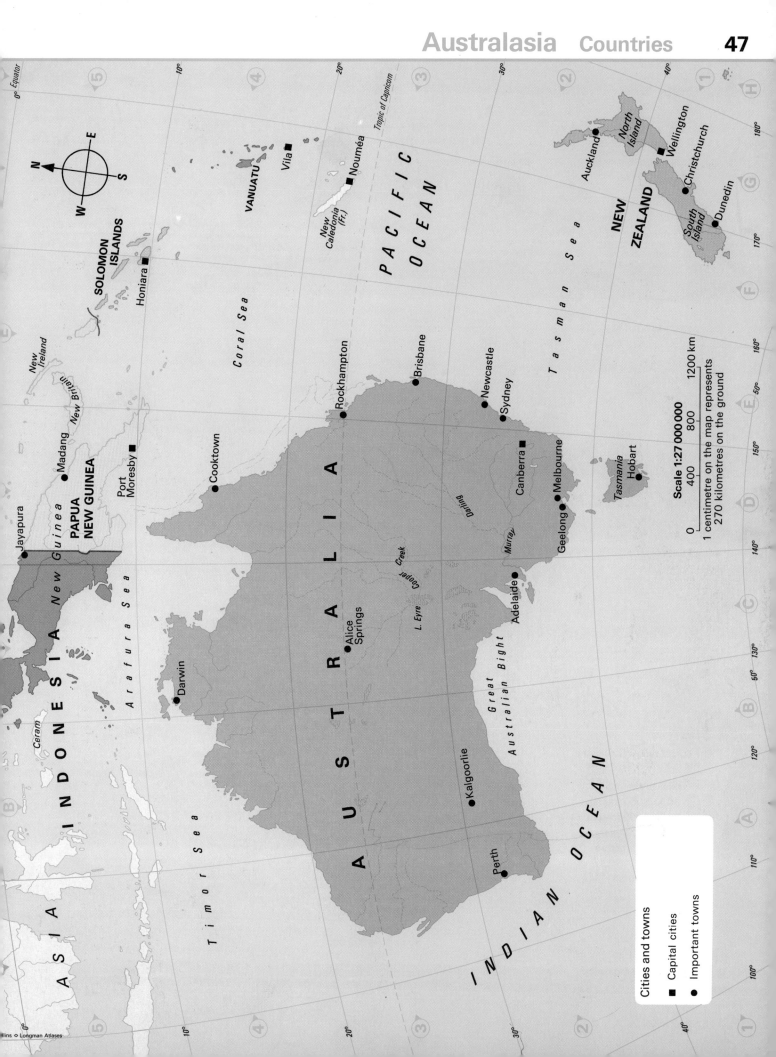

ASIA

INDONESIA

Jayapura

PAPUA NEW GUINEA

Madang

Port Moresby

New Guinea

New Ireland

New Britain

SOLOMON ISLANDS

Honiara

VANUATU

Vila

New Caledonia (Fr.)

Nouméa

Coral Sea

Arafura Sea

Timor Sea

Ceram

Equator

Tropic of Capricorn

PACIFIC OCEAN

NEW ZEALAND

North Island

South Island

Auckland

Wellington

Christchurch

Dunedin

Tasman Sea

AUSTRALIA

Darwin

Cooktown

Rockhampton

Brisbane

Newcastle

Sydney

Canberra

Alice Springs

Kalgoorlie

Perth

Adelaide

Geelong

Melbourne

Tasmania

Hobart

L. Eyre

Cooper Creek

Darling

Murray

Great Australian Bight

INDIAN OCEAN

N E S W

Scale 1:27 000 000

0 400 800 1200 km

1 centimetre on the map represents 270 kilometres on the ground

Cities and towns

■ Capital cities

● Important towns

Collins ◇ Longman Atlases

The colours show the height
of the land above sea level

- Over 5000 metres
- 3000-5000 metres
- 2000-3000 metres
- 1000-2000 metres
- 500-1000 metres
- 200-500 metres
- 0-200 metres

▲ 123 Mountain peak
(height in metres)

River

Lake

Land covered by ice

Scale 1:40 000 000

0   500   1000   1500   2000 km

1 centimetre on the map represents
400 kilometres on the ground

© Collins ◇ Longman Atlases

N W E S

ASIA
ARCTIC OCEAN
Bering Strait
Brooks Range
Yukon
Alaska Range
6194 Mt McKinley
Alaska Pen.
Kodiak I.
Banks I.
Victoria I.
Great Bear Lake
Mackenzie
Great Slave Lake
Mt Logan 6050
Peace
Coast Mountains
Vancouver I.
3954
Churchill
Saskatchewan
Nelson
Severn
Albany
Canadian Shield
Hudson Bay
Baffin Bay
Baffin Island
Labrador
Davis Strait
Ellesmere I.
GREENLAND
EUROPE
Arctic Circle
Newfoundland
C. Breton I.
ROCKY MOUNTAINS
Columbia
Coast Range
Great Salt Lake
Great Basin
Platte
Missouri
Colorado Plateau
Mt Elbert 4399
Colorado
Guadalupe I.
Gulf of California
Sierra Madre Occidental
Altiplano Mexicano
Sierra Madre Oriental
Rio Grande
5699
Sierra Madre
L. Nicaragua
Isthmus of Panama
Ozark Plateau
Arkansas
Red
Mississippi
Ohio
Tennessee
Appalachian Mts
2037
L. Superior
L. Michigan
L. Huron
St. Lawrence
L. Ontario
L. Erie
C. Cod
C. Hatteras
ATLANTIC OCEAN
Bermuda
Florida
Gulf of Mexico
Yucatan Peninsula
Cuba
Hispaniola
Greater Antilles
Lesser Antilles
Caribbean Sea
Bahama Islands
Tropic of Can
PACIFIC OCEAN
SOUTH AMERICA

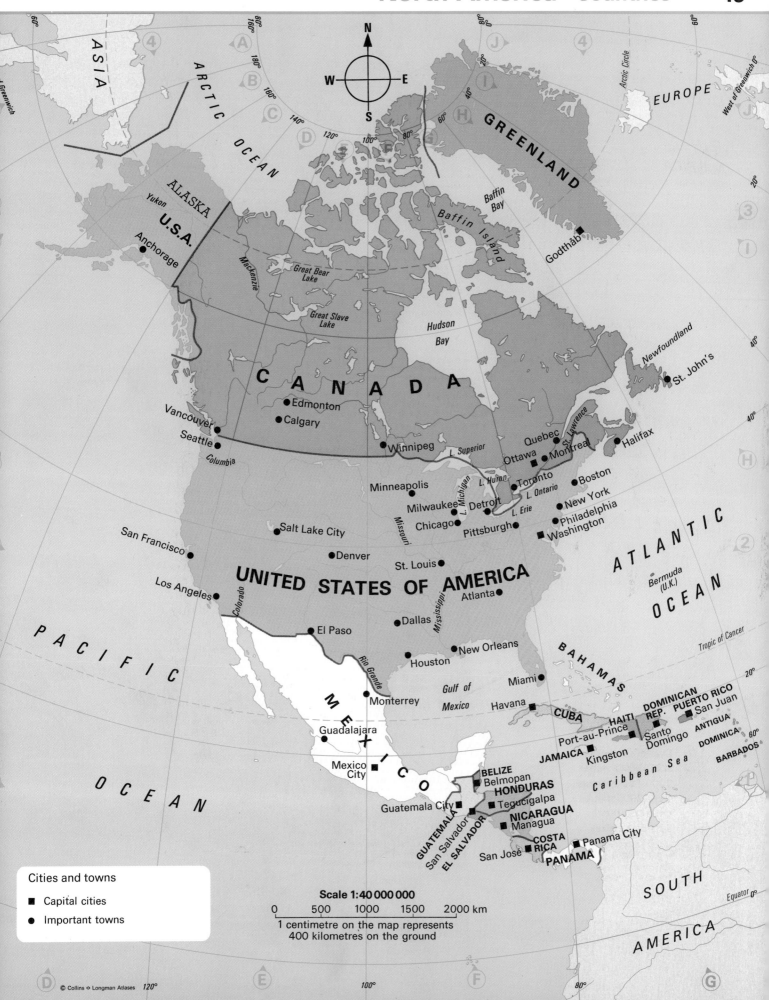

North America Countries

Cities and towns

- ■ Capital cities
- ● Important towns

Scale 1:40 000 000

0 500 1000 1500 2000 km

1 centimetre on the map represents
400 kilometres on the ground

ASIA

ARCTIC OCEAN

GREENLAND

Baffin Bay

Baffin Island

Godthåb

ALASKA
Yukon
U.S.A.

Anchorage

Mackenzie

Great Bear Lake

Great Slave Lake

Hudson Bay

Newfoundland

St. John's

C A N A D A

Edmonton
Calgary

Vancouver
Seattle
Columbia

Winnipeg
L. Superior

Quebec
St. Lawrence
Ottawa Montreal
Halifax

Minneapolis

L. Huron
Milwaukee L. Michigan Detroit
Missouri
Chicago
Pittsburgh

Toronto
L. Ontario
L. Erie
Boston
New York
Philadelphia
Washington

Salt Lake City

San Francisco

Denver
St. Louis

ATLANTIC

Los Angeles
Colorado

UNITED STATES OF AMERICA

Atlanta

Bermuda (U.K.)

OCEAN

El Paso
Rio Grande

Dallas

New Orleans

Houston

PACIFIC

M E X I C O

Monterrey

Gulf of Mexico

Miami

BAHAMAS

Tropic of Cancer

Havana
CUBA

DOMINICAN
REP. PUERTO RICO
San Juan

OCEAN

Guadalajara

HAITI
Port-au-Prince
JAMAICA
Kingston
Santo
Domingo
ANTIGUA

DOMINICA

Mexico City

BELIZE
Belmopan
HONDURAS
Guatemala City Tegucigalpa
GUATEMALA
San Salvador
EL SALVADOR
NICARAGUA
Managua
COSTA
RICA
San José
PANAMA
Panama City

BARBADOS

Caribbean Sea

SOUTH

AMERICA

Equator 0°

© Collins ○ Longman Atlases 120°

N
W · E
S

100° 80° 60° 40°

A B C D

6

Tropic of Cancer

20°

CENTRAL AMERICA

Caribbean Sea

ATLANTIC OCEAN

C. Gallinas Curaçao

Trinidad

L. Maracaibo

Llanos Orinoco

Mt. Roraima 2810

Guiana Highlands

5

Cordillera Occidental
Cordillera Central
Cordillera Oriental

Galapagos Is.

Negro Amazon

Equator 0°

Mt. Chimborazo 6272
Amazon

C. São Roque

C. Negra

S e l v a s

Madeira Tapajós

Tocantins

São Francisco

A N D E S

L. Titicaca 6550

Mato Grosso Upland

East Brazilian Highlands

4

Atacama Desert

L. Poopo

PACIFIC OCEAN

Gran Chaco

Paraguay

Paraná

Uruguay

C. Frio

20° Tropic of Capricorn

Mt. Aconcagua 6960

Pampas

Rio de la Plata

ATLANTIC OCEAN

Juan Fernandez Is.

3

A N D E S

Patagonia

Falkland Is.

S. Georgia

The colours show the height of the land above sea level.

Over 5000 metres
3000-5000 metres
2000-3000 metres
1000-2000 metres
500-1000 metres
200-500 metres
0-200 metres

123 Mountain peak
▲ (height in metres)

River

Lake

Tierra del Fuego

Magellan's Str. C. Horn

2

Scale 1:40 000 000

0   500   1000   1500   2000

1 centimetre on the map represents
400 kilometres on the ground

120°

Antarctic Peninsula

S. Orkney Is.

A B C D

120° 60° 100° 80° 60° 40° 20° 60°

1

0° © Collins · Longman

N
W E
S

CENTRAL
AMERICA

Caribbean Sea

Curaçao (Neth.)

ATLANTIC

OCEAN

Tropic of Cancer

Barranquilla
Caracas
TRINIDAD
AND TOBAGO
Maracaibo
Orinoco
Bucaramanga
VENEZUELA
GUYANA
Georgetown
Paramaribo
Medellin
SURINAM
Cayenne
Bogotá
GUIANA
(Fr.)
COLOMBIA
Cali
Quito
Galapagos Is.
(Ec.)
ECUADOR
Manaus
Amazon
Belém
São Luís
Equator 0°
Guayaquil
Iquitos
Fortaleza
Natal
P
B R A Z I L
E
Recife
R
Trujillo
São Francisco
Aracaju
U
Salvador
Lima
L. Titicaca
Brasília
La Paz
Arequipa
BOLIVIA
Belo Horizonte
PACIFIC
Sucre
Rio de Janeiro
São Paulo
Niterói
OCEAN
PARAGUAY
Santos
Antofagasta
Asunción
Curitiba
20°
Tropic of Capricorn
CHILE
Paraná
Uruguay
Pôrto Alegre
Córdoba
Juan Fernández Is.
(Chile)
Mendoza
URUGUAY
Valparaíso
Montevideo
ATLANTIC
Santiago
Buenos
Aires
Concepción
Mar del Plata
OCEAN
A
R
G
E
N
T
I
N
A
Falkland Is.
(U.K.)
Punta
Arenas
Tierra del
Fuego
S. Georgia
(U.K.)

Cities and towns

■ Capital cities

● Important towns

Scale 1:40 000 000

0   500   1000   1500   2000 km

1 centimetre on the map represents
400 kilometres on the ground

Antarctic
Peninsula

ns ◇ Longman Atlases 120°

The colours show the height of the land above sea level.

- Over 5000 metres
- 3000-5000 metres
- 2000-3000 metres
- 1000-2000 metres
- 500-1000 metres
- 200-500 metres
- 0-200 metres
- Land below sea level

ARCTIC OCEAN

West of Greenwic

Ellesmere Island

Greenland

Victoria Island

Baffin Bay

Arctic Circle

Baffin Island

Iceland

Yukon

Mackenzie

▲6194 Mt. McKinley

60°

Hudson Bay

British Isles

NORTH AMERICA

Mts. Missouri

Rocky

Great Lakes

St. Lawrence

Newfoundland

Colorado

Rio Grande

Mississippi

Appalachian Mts.

ATLANTIC

Atlas Mts

Tropic of Cancer

Gulf of Mexico

Canary Is.

Hawaiian Islands

20°

Cuba

S a h

A

Cape Verde Is.

Senegal

PACIFIC

Caribbean Sea

OCEAN

Futa Jalon

4

Orinoco

Gulf Guin

Equator

SOUTH

Amazon

Tocantins

3

AMERICA

OCEAN

Tuamotu Archipelago

Andes

Paraguay

20°

Tropic of Capricorn

▲6960 Mt. Aconcagua

2

Tierra del Fuego

60°

Antarctic Peninsula

Antarctic Circle

Weddell Sea

1

▲ Vinson Massif 5140

ANTA

A

160°

120°

B

80°

C

40°

D

HIGHEST MOUNTAINS

metres

Mt Everest 8848

Aconcagua 6960

Mt McKinley 6194

Kilimanjaro 5895

Vinson Massif 5140

Puntjak Jaya 5030

Mont Blanc 4807

5000

3000

2000

1000

500

200

| Asia | South America | North America | Africa | Antarctica | Australasia | Europe |

© Collins ◇ Longman Atlases

CONTINENT AREAS

North America 24 247 038 sq km

South America 17 821 028 sq km

Europe 10 354 636 sq km

Africa 30 244 049 sq km

of Greenwich
40°   80°   120°   160°

ARCTIC   OCEAN

pitsbergen
Barents   Sea

avia

 be A   Ob   Siberia   ASIA

URAL Mountains   Yenisei   Lena
Ural Mountains   Ob   Irtysh

Bering Sea
Sea of Okhotsk
60°

UROPE   Volga   Aral Sea   Amu Darya
Danube   Caucasus Mts.   Syr Darya   Caspian Sea
Black Sea   Dnieper

Sea of Japan
Honshu   5

nean Sea   Tigris   Euphrates   The Gulf   Arabia   Kunlun Shan   Huang He   Chang Jiang   Yellow Sea
A r a b i a   Himalaya   Brahmaputra   Mt. Everest   East China Sea
ICA   Indus   ▲ 8848   Yunnan Plateau
Ganges   Salween   PACIFIC   Tropic of Cancer   20°

a   Nile   Red Sea   Arabian Sea   Deccan   Bay of Bengal   Mekong   South China Sea
Chad   Gulf of Aden
Ethiopian Highlands   Sri Lanka   Philippines   Caroline Is.   Marshall Is.   4
Ubangi
Equator
Lake Victoria   ▲ 5895   INDIAN   Sumatra   Borneo   Puntjak Jaya   New Guinea   Solomon Is.
Kilimanjaro   ▲ 5030
Great Rift Valley   Java   3
Zambezi   OCEAN   Timor Sea   Coral Sea   Fiji Is.
Madagascar   AUSTRALASIA
Limpopo   Great Sandy Desert   20°
Orange   Australia   Tropic of Capricorn
L. Eyre
Darling   Great Dividing Range   Tasman Sea   2
Murray   Tasmania   New Zealand

SOUTHERN   OCEAN
60°

Antarctic Circle

Ross Sea   1

TICA

40°   80°   120°   160°
F   G   H   I

Asia   44 391 162 sq km
Australasia   8 547 000 sq km
Antarctica   13 338 500 sq km

☐ Land covered by ice
▲ 4807   Mountain peak (height in metres)
⬛ Lake
〜 River

Scale 1:85 000 000
0   1000   2000   3000 km
1 centimetre on the map represents
850 kilometres on the ground

LONGEST RIVERS

River	Length	Region
Nile	6695	Africa
Amazon	6570	South America
Mississippi-Missouri	6020	North America
Chang Jiang (Yangtze)	5471	Asia
Murray-Darling	3717	Australasia
Volga	3688	Europe

kilometres   7000   6000   5000   4000   3000   2000   1000   0

© Collins ◇ Longman Atlases

Cities and towns

- ■ Capital cities
- • Important towns

A. : ALBANIA
AU. : AUSTRIA
B. : BELGIUM
BE. : BENIN
BU. : BURKINA
C. : CZECHOSLOVAKIA
CAM. : CAMEROON
C.A.R. : CENTRAL AFRICAN
REPUBLIC
E.G. : EAST GERMANY
EQ.G. : EQUATORIAL
GUINEA
G. : GAMBIA
G.B. : GUINEA BISSAU
GH. : GHANA
I.C. : IVORY COAST
L. : LUXEMBOURG
LE. : LESOTHO
N. : NETHERLANDS
R. : RWANDA
S. : SWITZERLAND
SEN. : SENEGAL
S.L. : SIERRA LEONE
SW. : SWAZILAND
T. : TOGO
U.A.E. : UNITED ARAB
EMIRATES
W.G. : WEST GERMANY
Y. : YUGOSLAVIA
Z. : ZIMBABWE

The view of the world on the large map and on Map A is the one you can probably recognise straight away. It's the view you are used to, with Britain and Europe right in the centre.

But the world does not have to be shown this way. People from other parts of the world find it more convenient to draw world maps in a different way. Map B and Map C are two examples of world maps that are not *eurocentric,* meaning centred on Europe.

Map A

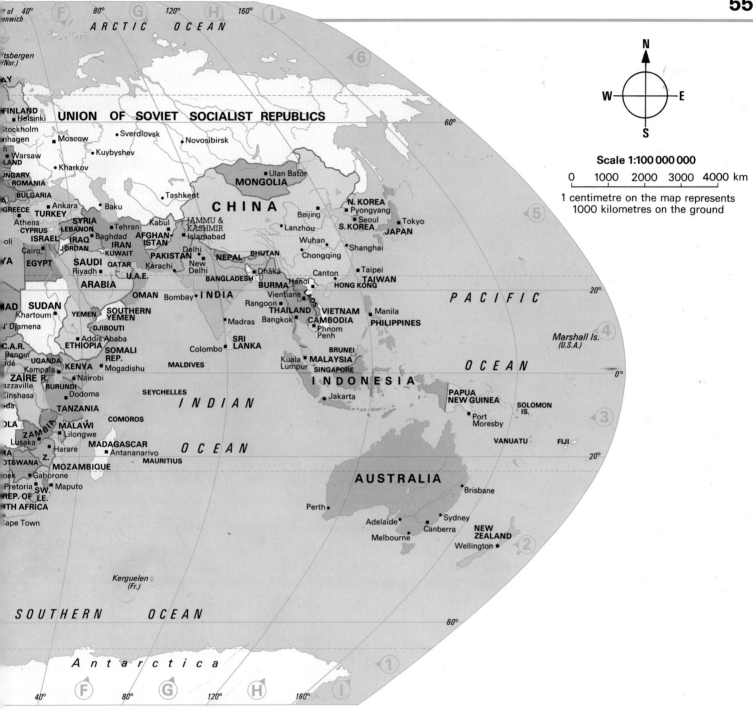

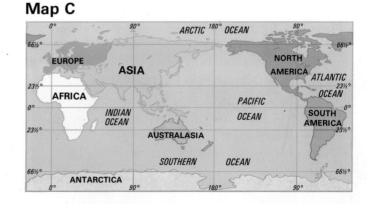

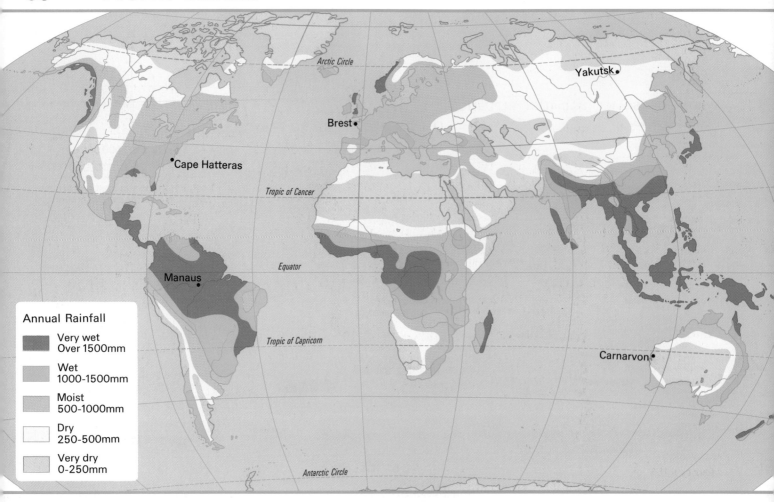

Annual Rainfall

	Very wet Over 1500mm
	Wet 1000-1500mm
	Moist 500-1000mm
	Dry 250-500mm
	Very dry 0-250mm

CLIMATE GRAPHS

Average monthly rainfall

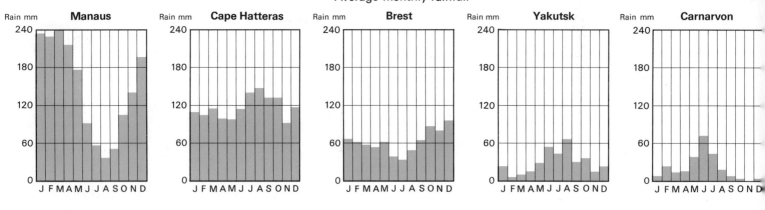

Manaus — Rain mm

Cape Hatteras — Rain mm

Brest — Rain mm

Yakutsk — Rain mm

Carnarvon — Rain mm

A desert in a very dry, hot region

A forest in a very wet, hot region

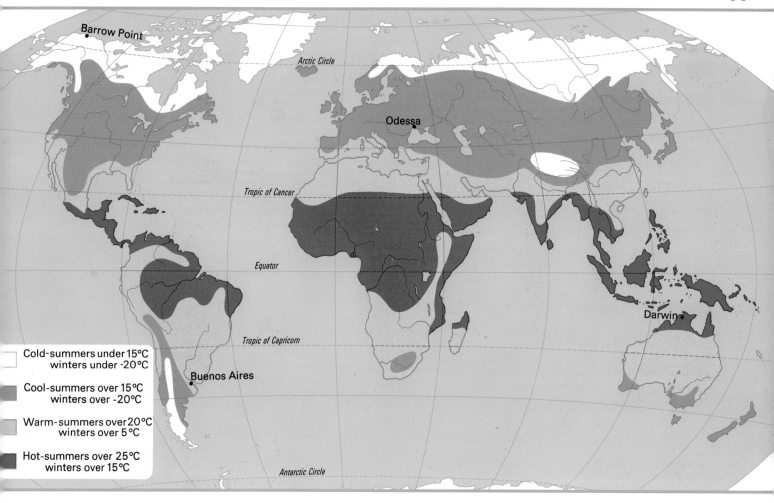

Cold-summers under 15°C
winters under -20°C

Cool-summers over 15°C
winters over -20°C

Warm-summers over 20°C
winters over 5°C

Hot-summers over 25°C
winters over 15°C

CLIMATE GRAPHS

Average monthly temperature

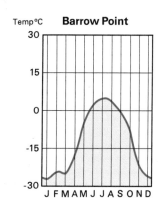

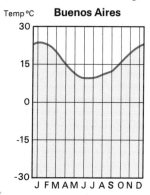

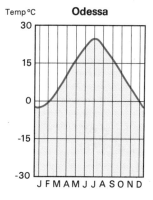

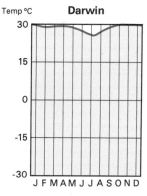

A glacier in a very dry, cold region

Grassland in a moist, cool region

Population

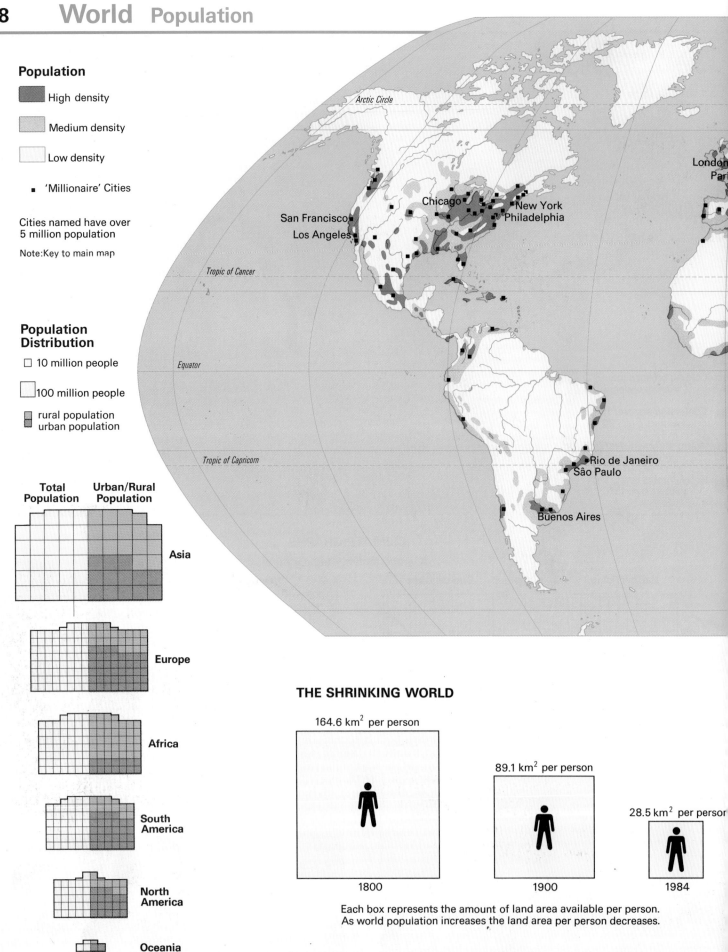

- ▨ High density
- ▨ Medium density
- ☐ Low density
- ■ 'Millionaire' Cities

Cities named have over
5 million population

Note: Key to main map

Population Distribution

- ☐ 10 million people
- ☐ 100 million people
- ▨ rural population
  urban population

Total Population	Urban/Rural Population	
		Asia
		Europe
		Africa
		South America
		North America
		Oceania

Map labels: Arctic Circle, Chicago, New York, Philadelphia, San Francisco, Los Angeles, Tropic of Cancer, Equator, Tropic of Capricorn, Rio de Janeiro, São Paulo, Buenos Aires, London, Paris

THE SHRINKING WORLD

164.6 km² per person — 1800

89.1 km² per person — 1900

28.5 km² per person — 1984

Each box represents the amount of land area available per person.
As world population increases the land area per person decreases.

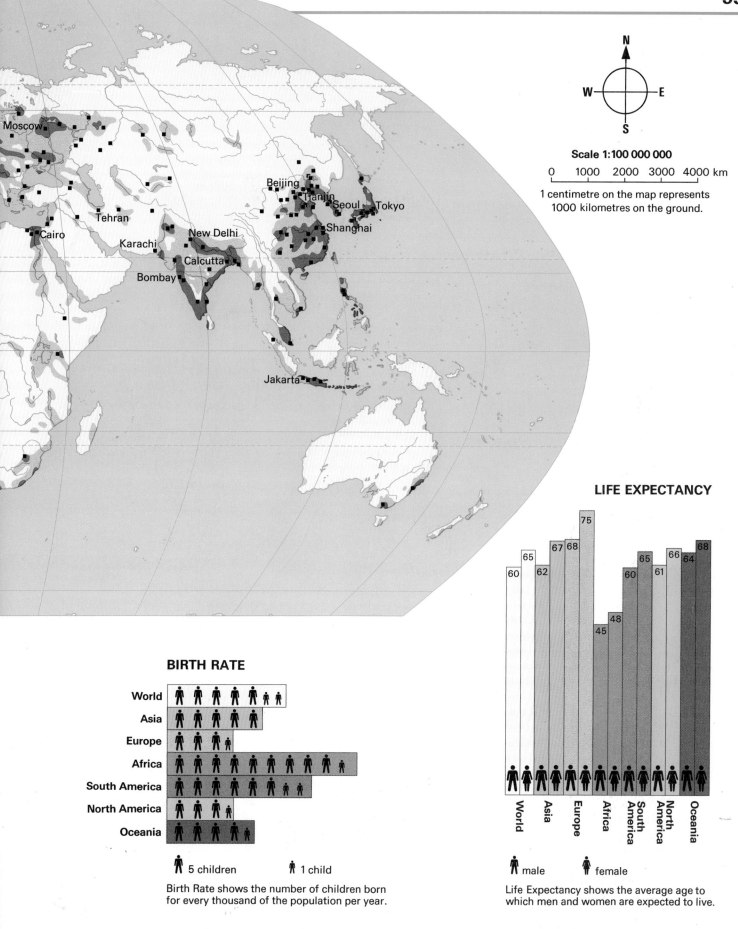

N
W — E
S

Scale 1:100 000 000

0   1000   2000   3000   4000 km

1 centimetre on the map represents
1000 kilometres on the ground.

Moscow
Tehran
Cairo
Karachi
New Delhi
Calcutta
Bombay
Beijing
Tianjin
Seoul
Tokyo
Shanghai
Jakarta

LIFE EXPECTANCY

World	Asia	Europe	Africa	South America	North America	Oceania
60 / 65	62 / 67	68 / 75	45 / 48	60 / 65	61 / 66	64 / 68

male   female

Life Expectancy shows the average age to
which men and women are expected to live.

BIRTH RATE

World
Asia
Europe
Africa
South America
North America
Oceania

5 children   1 child

Birth Rate shows the number of children born
for every thousand of the population per year.

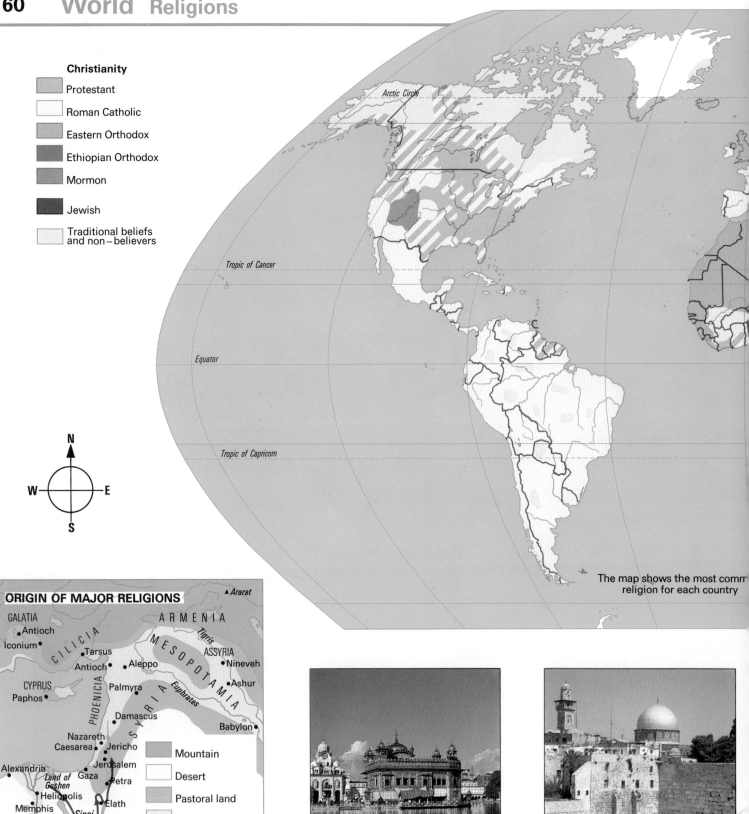

Christianity

Protestant

Roman Catholic

Eastern Orthodox

Ethiopian Orthodox

Mormon

Jewish

Traditional beliefs and non−believers

N
W E
S

The map shows the most comm religion for each country

ORIGIN OF MAJOR RELIGIONS

▲ Ararat

GALATIA
• Antioch
ARMENIA
CILICIA
MESOPOTAMIA
Tigris
Iconium •
Tarsus •
ASSYRIA
Antioch •
Aleppo •
Nineveh •
CYPRUS
PHOENICIA
Palmyra •
Ashur •
Paphos •
SYRIA
Euphrates
Damascus •
Babylon •
Nazareth •
Caesarea •
Jericho •
Jerusalem •
Alexandria •
Gaza •
Petra •
Land of Goshen
Heliopolis •
Elath •
Memphis •
Sinai
Mt. Sinai
EGYPT
Nile
SINUS ARABICUS
ARABIA
Abydos •
• Thebes
• Medina
Syene •
• Mecca

Mountain

Desert

Pastoral land

Fertile land

• Important city

→ Route of the exodus

Jerusalem – centre of Christianity and Judaism

Mecca – centre of Islamic world

Golden Temple of the Sikhs in Amritsar, India

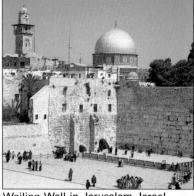

Wailing Wall in Jerusalem, Israel. Jews worship here.

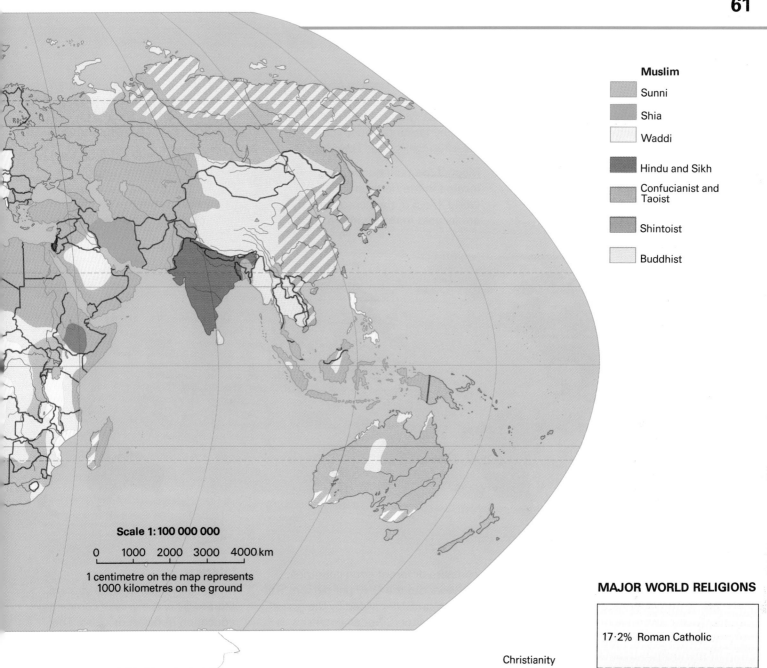

Muslim

- Sunni
- Shia
- Waddi

- Hindu and Sikh
- Confucianist and Taoist
- Shintoist
- Buddhist

Scale 1:100 000 000

0   1000 2000 3000 4000 km

1 centimetre on the map represents
1000 kilometres on the ground

Mohammed Ali Mosque in Cairo,
Egypt. Muslims worship here.

Religions in Britain — 1985

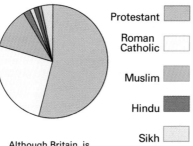

- Protestant
- Roman Catholic
- Muslim
- Hindu
- Sikh
- Jewish
- Mormon
- Other

Although Britain is often called a 'Christian nation' only 15% of the total adult population are members of a church. The diagram shows only these church members

MAJOR WORLD RELIGIONS

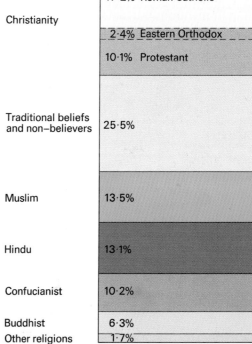

Christianity	17·2% Roman Catholic
	2·4% Eastern Orthodox
	10·1% Protestant
Traditional beliefs and non–believers	25·5%
Muslim	13·5%
Hindu	13·1%
Confucianist	10·2%
Buddhist	6·3%
Other religions	1·7%

Some countries in the world are much richer than others. Great differences in wealth exist between countries, and between the people within any one country.

The richer or **developed** countries have built up their industries to a high level. This gives most of the people a good standard of living.

The poor or **developing** countries are still building up their industries. Many people make their living by farming. These countries are often called the **'Third World'**.

Most of the richer countries are located in the northern part of the world. Most of the poorer countries are located in the southern parts

of the world. The richer countries are sometimes grouped together and called **The North** and the poorer countries are called **The South**.

These maps show just four ways of looking at living standards in different countries. But there are many things that can be used - there is no one way of comparing 'wealth'.

G.N.P. PER PERSON

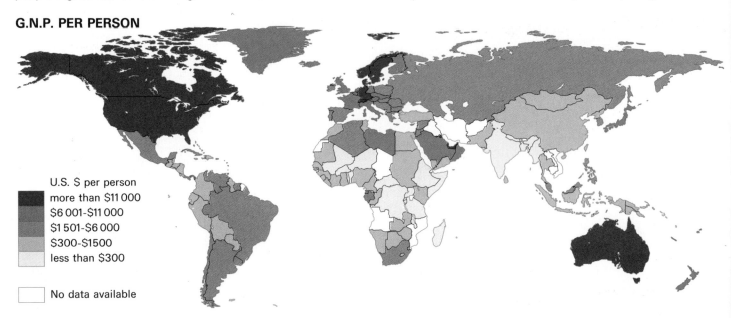

U.S. $ per person
more than $11 000
$6 001-$11 000
$1 501-$6 000
$300-$1500
less than $300

No data available

G.N.P. per person

The total amount of money a country earns from its industry and services both at home and abroad in one year is called its Gross National Product (or G.N.P.). To compare different sized countries we work out the GNP per person. This is the total GNP divided by the total population. Not all countries measure their money in the same way, so on the map GNP is shown in US$.

Literacy

If you can read and write you are literate. This is a way of comparing education in different areas. The map shows the percentage of people over the age of 15 who are literate.

LITERACY

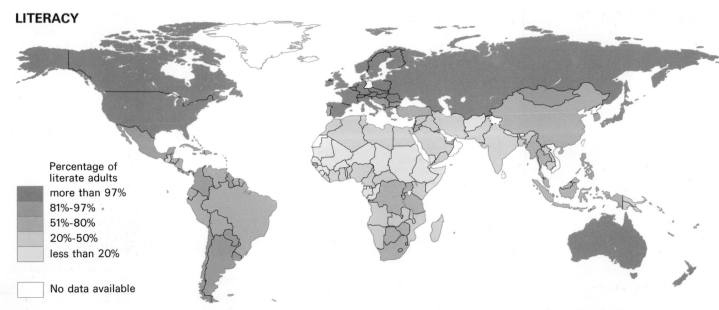

Percentage of literate adults
more than 97%
81%-97%
51%-80%
20%-50%
less than 20%

No data available

HOW IMPORTANT IS FARMING

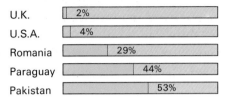

U.K. 2%
U.S.A. 4%
Romania 29%
Paraguay 44%
Pakistan 53%

The percentage of each country's total labour force that was employed in farming in 1983

HOW MUCH IS SPENT ON EDUCATION

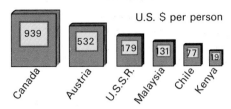

U.S. $ per person

939 Canada
532 Austria
179 U.S.S.R.
131 Malaysia
77 Chile
19 Kenya

The amount each country spent on education divided by the total population in 1982

HOW MUCH AID IS GIVEN TO OTHER COUNTRIES

U.S. $ per person

966 Kuwait
138 Norway
53 Australia
34 U.K.
33 U.S.A.
U.S.S.R.

The amount each country gave as development aid divided by the total population in 1983

MEDICAL CARE

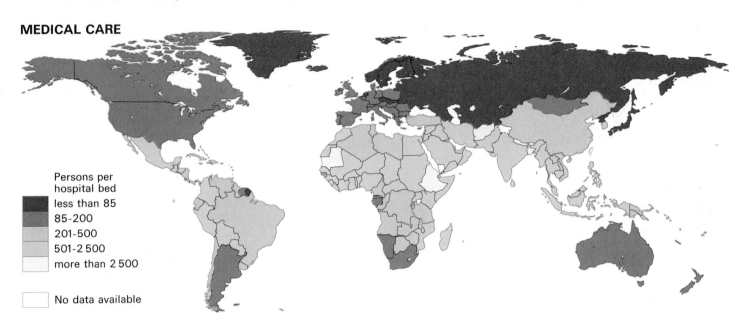

Persons per hospital bed
- less than 85
- 85-200
- 201-500
- 501-2 500
- more than 2 500

No data available

Medical Care

How many people in a country have the chance of good medical care when they are sick? This is one way we can look at how rich or developed a country has become.

There are a number of ways of measuring medical care. One way is to look at how many hospital beds the country has compared to the total number of people who live there.

Infant Mortality

Some babies do not survive until their first birthday. We measure infant mortality, by counting, out of every 1000 babies born alive, how many die before the age of one year.

INFANT MORTALITY

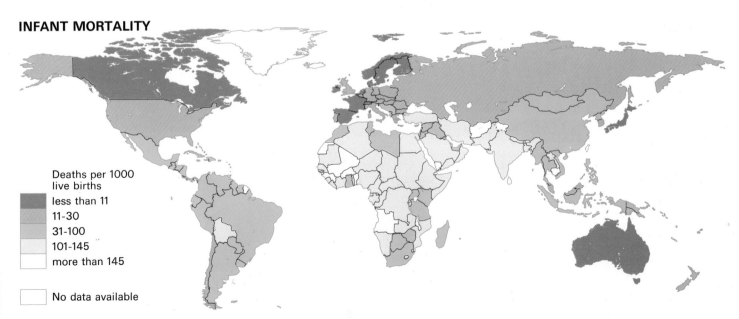

Deaths per 1000 live births
- less than 11
- 11-30
- 31-100
- 101-145
- more than 145

No data available

© Collins ◇ Longman Atlases

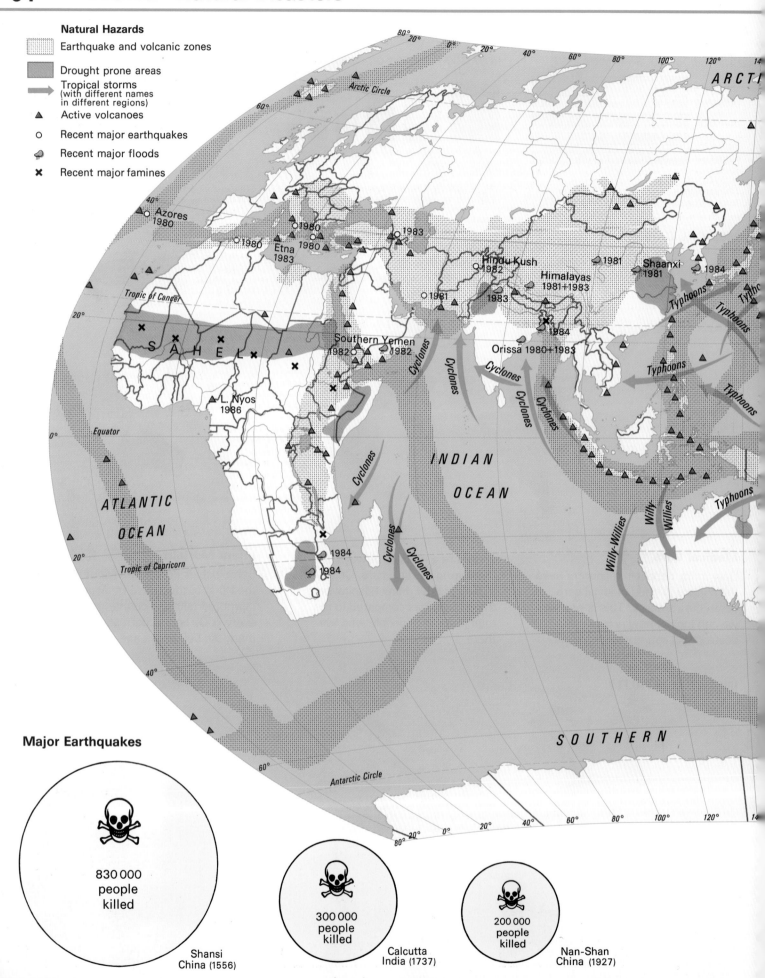

Natural Hazards

Earthquake and volcanic zones

Drought prone areas

→ Tropical storms (with different names in different regions)

▲ Active volcanoes

○ Recent major earthquakes

Recent major floods

✕ Recent major famines

Major Earthquakes

830 000 people killed
Shansi China (1556)

300 000 people killed
Calcutta India (1737)

200 000 people killed
Nan-Shan China (1927)

© Collins ◇ Longman Atlases

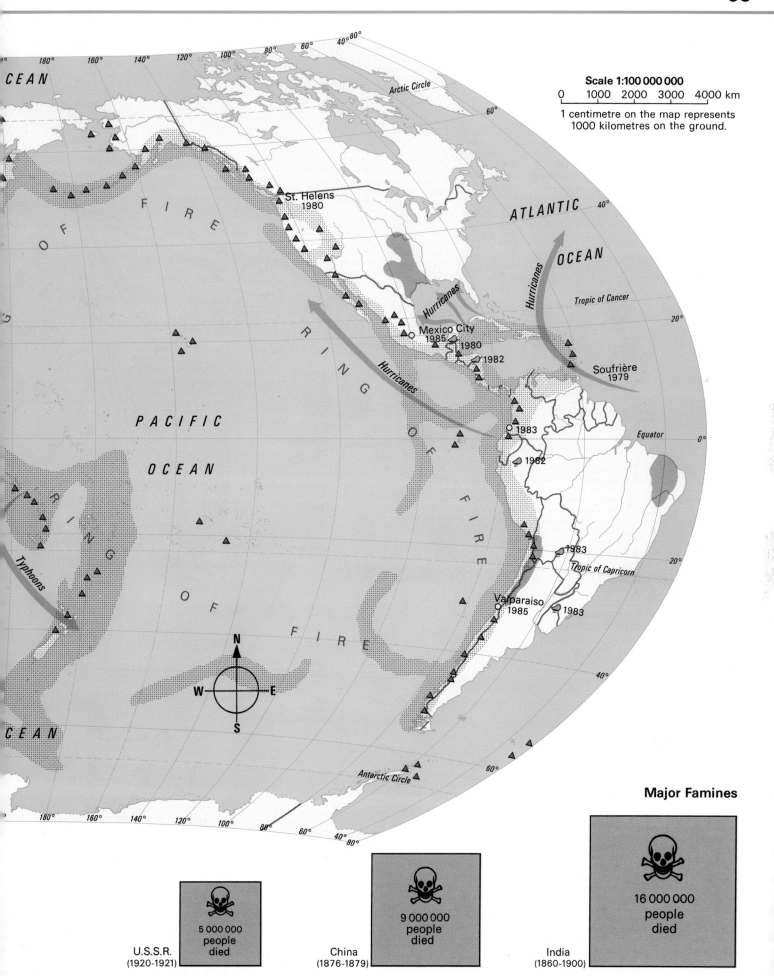

Scale 1:100 000 000

0   1000 2000 3000 4000 km

1 centimetre on the map represents
1000 kilometres on the ground.

CEAN

Arctic Circle

ATLANTIC

St. Helens
1980

OCEAN

40°

Hurricanes

Tropic of Cancer

Mexico City
1985

Hurricanes

1980

20°

1982

Soufrière
1979

PACIFIC

1983

Equator 0°

OCEAN

1982

RING OF FIRE

Typhoons

1983

Tropic of Capricorn 20°

Valparaiso
1985

1983

N

40°

W   E

S

CEAN

Antarctic Circle

60°

Major Famines

U.S.S.R.
(1920-1921)

5 000 000
people
died

China
(1876-1879)

9 000 000
people
died

India
(1860-1900)

16 000 000
people
died

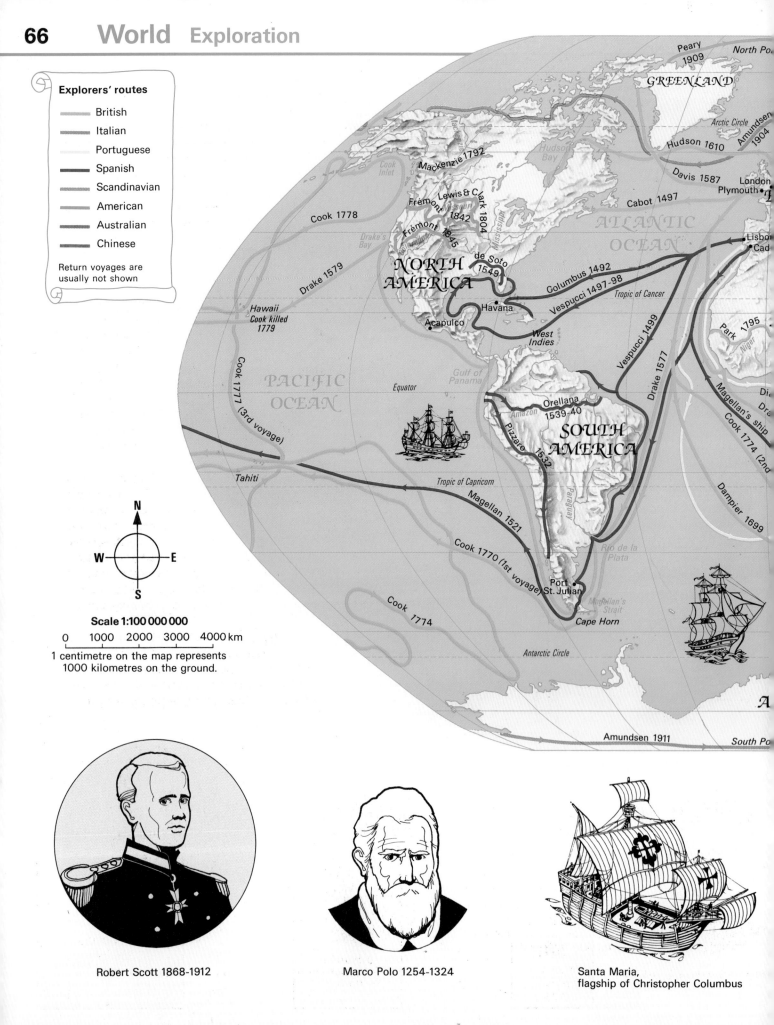

Explorers' routes

—	British
—	Italian
—	Portuguese
—	Spanish
—	Scandinavian
—	American
—	Australian
—	Chinese

Return voyages are usually not shown

N W E S

Scale 1:100 000 000

0   1000   2000   3000   4000 km

1 centimetre on the map represents 1000 kilometres on the ground.

Peary 1909
North Po
GREENLAND
Arctic Circle
Amundsen 1904
Hudson 1610
Davis 1587
London
Plymouth
Cook Inlet
Mackenzie 1792
Cabot 1497
ATLANTIC OCEAN
Lisbo
Cad
Frémont 1842
Lewis & Clark 1804
Frémont 1845
Cook 1778
Drake's Bay
de Soto 1549
NORTH AMERICA
Drake 1579
Columbus 1492
Vespucci 1497-98
Tropic of Cancer
Hawaii
Cook killed 1779
Havana
Acapulco
West Indies
Vespucci 1499
Park 1795
Niger
PACIFIC OCEAN
Equator
Gulf of Panama
Drake 1577
Magellan's ship
Die
Dra
Cook 1777 (3rd voyage)
Orellana 1539-40
Amazon
SOUTH AMERICA
Cook 1774 (2
Pizarro 1532
Dampier 1699
Tahiti
Tropic of Capricorn
Magellan 1521
Paraguay
Cook 1770 (1st voyage)
Rio de la Plata
Port St. Julian
Magellan's Strait
Cook 1774
Cape Horn
Antarctic Circle
A
Amundsen 1911
South Po

Robert Scott 1868-1912

Marco Polo 1254-1324

Santa Maria, flagship of Christopher Columbus

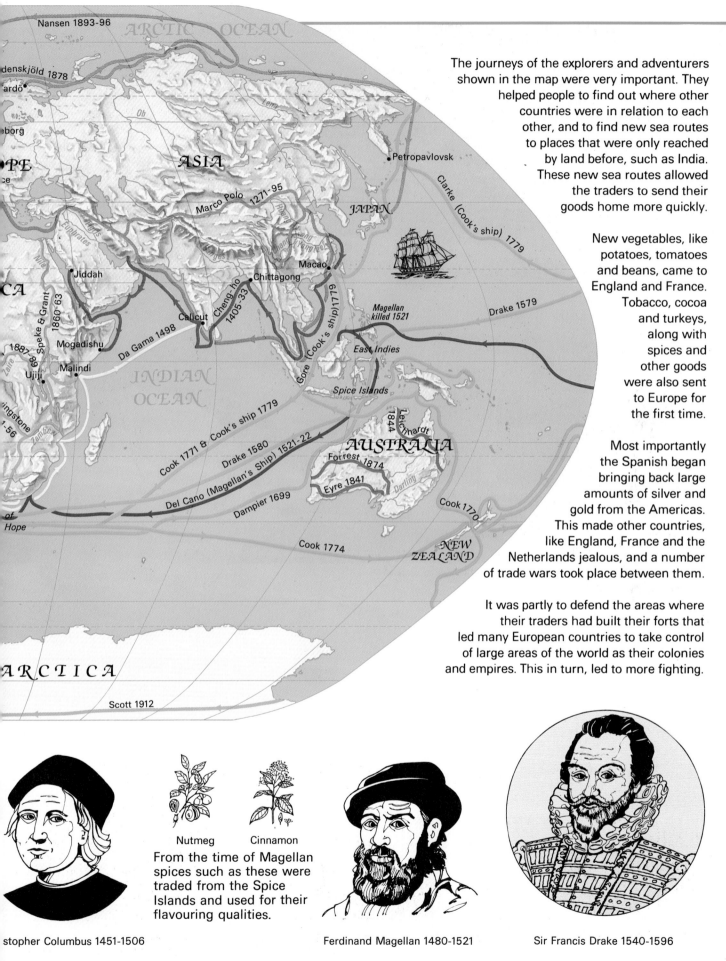

The journeys of the explorers and adventurers shown in the map were very important. They helped people to find out where other countries were in relation to each other, and to find new sea routes to places that were only reached by land before, such as India. These new sea routes allowed the traders to send their goods home more quickly.

New vegetables, like potatoes, tomatoes and beans, came to England and France. Tobacco, cocoa and turkeys, along with spices and other goods were also sent to Europe for the first time.

Most importantly the Spanish began bringing back large amounts of silver and gold from the Americas. This made other countries, like England, France and the Netherlands jealous, and a number of trade wars took place between them.

It was partly to defend the areas where their traders had built their forts that led many European countries to take control of large areas of the world as their colonies and empires. This in turn, led to more fighting.

Map labels:

Nansen 1893-96
ARCTIC OCEAN
denskjöld 1878
ardö
borg
ASIA
Petropavlovsk
Clarke (Cook's ship) 1779
Marco Polo 1271-95
JAPAN
Macao
Jiddah
Chittagong
Cheng-ho 1405-33
Calicut
Drake 1579
Da Gama 1498
Magellan killed 1521
Mogadishu
Gore (Cook's ship) 1779
Speke & Grant 1860-63
1887-89
East Indies
Zaire
Malindi
INDIAN OCEAN
Spice Islands
Ujiji
ingstone 1-56
Zambezi
Leichhardt 1844
Cook 1771 & Cook's ship 1779
AUSTRALIA
Drake 1580
Forrest 1874
Del Cano (Magellan's Ship) 1521-22
Eyre 1841
Darling
Dampier 1699
Cook 1770
of Hope
Cook 1774
NEW ZEALAND
ARCTICA
Scott 1912

Nutmeg   Cinnamon

From the time of Magellan spices such as these were traded from the Spice Islands and used for their flavouring qualities.

stopher Columbus 1451-1506

Ferdinand Magellan 1480-1521

Sir Francis Drake 1540-1596

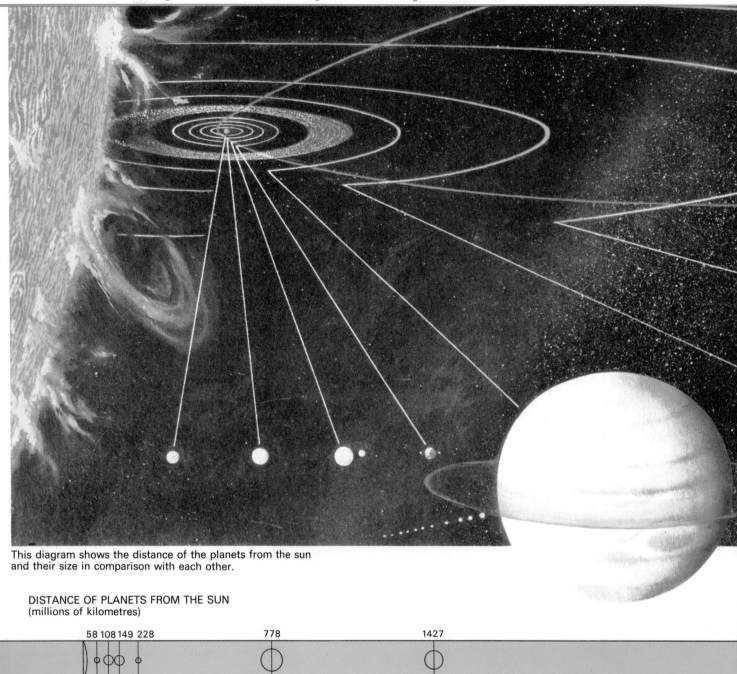

This diagram shows the distance of the planets from the sun and their size in comparison with each other.

DISTANCE OF PLANETS FROM THE SUN
(millions of kilometres)

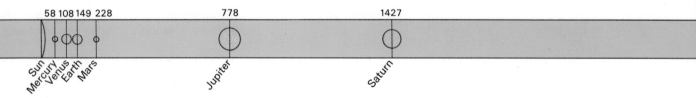

58 108 149 228   778   1427

Sun Mercury Venus Earth Mars   Jupiter   Saturn

	SUN	**MERCURY**	**VENUS**	**EARTH**	**MARS**	**JUPITER**	**SATURN**	**URANUS**	**NEPTUNE**	**PLUTO**
Distance from the sun (millions of kilometres)	–	58	108	149.5	228	778.5	1427	2870	4497	5900
Diameter in Kilometres	1 392 000	4850	12 104	12 756	6790	142 600	120 000	52 000	48 000	3000
Rotation period	25days9h	58days14h	243 days	23h 56m	24h 37m	9h 50m	10h 14m	24 hours	22 hours	6days9h
Surface temperature in °C	6000°	350°/-170°	480°	22°	-50°	-150°	-180°	-210°	-220°	-230°
Size (Earth = 1)	333 434·00	0.04	0.83	1.00	0.11	318.00	95.00	15.00	17.00	0.06
Length of year	–	88 days	225 days	1 year	1yr 322d	11yrs315d	29yrs167d	84 yrs 6d	164yrs288d	247yrs255d
Number of satellites	–	0	0	1	2	14	17	5	2	1

4497
Neptune

5900
Pluto

SPACE EXPLORATION

1957 Launch of Sputnik 1 by USSR –the first man-made Earth satellite.
1959 Lunik 2 hits the Moon – the first object to reach another world.
1961 The first man in space, Yuri Gagarin, orbits the earth once in Vostok 1.
1967 Venera 3 makes a successful landing on Venus.
1969 American astronaut, Neil Armstrong from Apollo 11 becomes the first man to set foot on the Moon.

1973 Pioneer 10 takes closeup photos of Jupiter.
1974 Mariner 10 takes closeup photos of Mercury.
1976 First successful landings on Mars by two Viking spacecraft.
1979 Pioneer 11 takes closeup photos of Saturn.
1981 American space shuttle, Columbia, makes its first test flight.
1982 Columbia is launched on its first operational flight.

1983 Dr Sally Ride becomes the first woman astronaut on Challenger, the American shuttle mission.
1985 the USSR launches a new type of space station called Mir, meaning "Peace".
1986 Challenger explodes shortly after lift-off with the loss of all seven members of the crew.
1986 Russian cosmonaut, Svetlana Savitskaya, from Soyuz T12 becomes the first woman to walk in space.

Index

Abbreviations

Afghan.	Afghanistan
Austa.	Australasia
b. or **B.**	bay
Bangla.	Bangladesh
c. or **C.**	cape
C.A.R.	Central African Republic
Cen. America	Central America
Ch. Is.	Channel Islands
Czech.	Czechoslovakia
d.	county or region
des.	desert
Dom. Rep.	Dominican Republic
E.	East
Equat. Guinea	Equatorial Guinea
est.	estuary
f.	physical feature
g. or **G.**	gulf
i. or **I.**	island
is. or **Is.**	islands
I.O.M.	Isle of Man
l. or **L.**	lake
Med. Sea	Mediterranean Sea
Mt.	mount
mtn. or **Mtn.**	mountain
mts. or **Mts.**	mountains
Neth.	Netherlands
N.	North
N. Ire.	Northern Ireland
Oc.	Ocean
P.N.G.	Papua New Guinea
pen. or **Pen.**	peninsula
Phil.	Philippines
Pt.	point
r.	river
Rep. of Ire.	Republic of Ireland
R.S.A.	Republic of South Africa
Somali Rep.	Somali Republic
S.	south or southern
str. or **Str.**	strait
Switz.	Switzerland
U.S.S.R	Union of Soviet Socialist Republics
U.K.	United Kingdom
U.S.A.	United States of America
W.	west or western
Yugo.	Yugoslavia

Aberdeen Scotland 22 **F4** 57N 2W
Aberystwyth Wales 21 **C4** 52N 4W
Abuja Nigeria 41 **B3** 9N 7E
Accra Ghana 41 **A3** 6N 0
Achill I. Rep. of Ire. 13 **A3** 53N 10W
Aconcagua, Mt. Argentina 50 **B2** 33S 70W
Addis Ababa Ethiopia 41 **C3** 9N 39E
Aden S. Yemen 45 **C2** 13N 45E
Aden, G. of Indian Oc. 44 **C2** 13N 50E
Adriatic Sea Med. Sea 32 **E2** 43N 16E
Aegean Sea Med. Sea 32 **F1** 39N 25E
Afghanistan Asia 45 **D3** 34N 65E
Agulhas, C. R.S.A. 40 **C1** 35S 20E
Ahaggar Mts. Algeria 40 **B4** 24N 6E
Aire *r.* England 13 **F3** 53N 1W
Alaska *d.* U.S.A. 49 **C4** 67N 150W
Alaska Pen. U.S.A. 48 **C3** 56N 160W
Alaska Range *mts.* U.S.A. 48 **C4** 62N 152W
Albania Europe 33 **E2** 41N 20E
Aldershot England 21 **G3** 51N 0
Algeria Africa 41 **B4** 28N 2E
Algiers Algeria 41 **B4** 37N 3E
Alloa Scotland 22 **E3** 56N 3W
Alps *mts.* Europe 32 **D2** 46N 8E
Altai, Mts. Mongolia 44 **F4** 46N 93E
Altiplano Mexicano *mts.* N. America 48 **E2** 24N 105W
Amazon *r.* Brazil 50 **C3** 2S 50W
Amman Jordan 45 **C3** 32N 36E
Amsterdam Neth. 33 **D3** 52N 5E
Amur *r.* U.S.S.R. 44 **H4** 53N 140E
Anchorage U.S.A. 49 **C4** 61N 150W
Andaman Is. India 44 **F2** 12N 93E
Andes *mts.* S. America 50 **B3** 15S 72W
Andorra Europe 33 **D2** 42N 1E
Andover England 21 **F3** 51N 1W
Anglesey *i.* Wales 13 **D3** 53N 4W
Angola Africa 41 **B2** 12S 18E
Ankara Turkey 45 **C3** 40N 33E
Annan Scotland 22 **E1** 54N 3W
Annan *r.* Scotland 13 **E4** 55N 3W
Antananarivo Madagascar 41 **D2** 19S 47E
Antigua Leeward Is. 49 **G1** 17N 62W
Antrim N. Ire. 23 **E4** 54N 6W
Antrim *d.* N. Ire. 23 **E4** 54N 6W
Antrim, Mts.of N. Ire. 13 **C4** 55N 6W
Apennines *mts.* Italy 32 **E2** 42N 13E
Appalachian Mts. U.S.A. 48 **G2** 40N 78W
Arabia Asia 44 **C3** 25N 45E
Arabian Sea Asia 44 **D2** 16N 65E
Arafura Sea Austa. 46 **C5** 9S 135E
Aral Sea U.S.S.R. 44 **D4** 45N 60E
Aran Is. Rep. of Ire. 13 **B3** 53N 9W
Ararat, Mt. Turkey 44 **C3** 40N 44E
Arbroath Scotland 22 **F3** 56N 2W
Arctic Ocean 53 **F6** 75N 65E
Ardennes *mts.* Belgium 32 **D2** 50N 5E
Ardrossan Scotland 22 **D2** 55N 4W
Argentina S. America 51 **B2** 35S 65W
Arkansas *r.* U.S.A. 48 **F2** 34N 91W
Armagh N. Ire. 23 **E4** 54N 6W
Armagh *d.* N. Ire. 23 **E4** 54N 6W
Arnhem Land *f.* Australia 46 **C4** 13S 133E
Arran *i.* Scotland 13 **D4** 55N 5W
Ashford England 21 **H3** 51N 0
Ashington England 20 **F7** 55N 1W
Asunción Paraguay 51 **C2** 25S 58W
Atacama Desert S. America 50 **B3** 20S 69W
Athens Greece 33 **F1** 38N 24E
Atlantic Ocean 52 **C5** 25N 40W
Atlas Mts. Africa 40 **A4** 33N 4W
Australia Austa. 47 **C3** 25S 135E
Austria Europe 33 **E2** 48N 14E
Avon *r.* England 13 **F3** 52N 1W
Avon *d.* England 21 **E3** 51N 2W
Awe, Loch Scotland 13 **D5** 56N 5W
Aylesbury England 21 **G3** 51N 0
Ayr Scotland 22 **D2** 55N 4W
Azov, Sea of U.S.S.R. 32 **G2** 46N 36E

Baffin B. Canada 48 **G4** 74N 70W
Baghdad Iraq 45 **C3** 33N 44E
Bahamas Cen. America 49 **G2** 24N 75W
Bahrain Asia 45 **D3** 26N 51E
Bahr el Jebel *r.* Sudan 40 **C3** 10N 30E
Baikal, L. U.S.S.R. 44 **F4** 53N 108E
Baku U.S.S.R. 45 **C4** 40N 50E
Balearic Is. Spain 32 **D1** 39N 2E
Balkan Mts. Bulgaria 32 **F2** 43N 25E
Balkhash, L. U.S.S.R. 44 **E4** 47N 75E
Ballymena N. Ire. 23 **E4** 54N 6W
Baltic Sea Europe 32 **E3** 56N 19E
Bamako Mali 41 **A3** 13N 8W
Banbury England 21 **F4** 52N 1W
Banff Scotland 22 **F4** 57N 2W
Bangkok Thailand 45 **F2** 14N 101E
Bangladesh Asia 45 **E3** 24N 90E
Bangor N. Ire. 23 **F4** 54N 5W
Bangor Wales 20 **C5** 53N 4W
Bangui C.A.R. 41 **B3** 4N 19E
Banjul Gambia 41 **A3** 13N 17W
Barbados Lesser Antilles 49 **G1** 13N 60W
Barcelona Spain 33 **D2** 41N 2E
Barkly Tableland *f.* Australia 46 **C4** 19S 137E
Barnsley England 20 **F5** 53N 1W
Barnstaple England 21 **C3** 51N 4W
Barra *i.* Scotland 13 **C5** 56N 7W
Barrow *r.* Rep. of Ire. 13 **C3** 52N 6W
Barrow-in-Furness England 20 **D6** 54N 3W
Barry Wales 21 **D3** 51N 3W
Basildon England 21 **H3** 51N 0
Basingstoke England 21 **F3** 51N 1W
Bass Str. Australia 46 **D2** 40S 146E
Bath England 21 **E3** 51N 2W
Bathgate Scotland 22 **E2** 55N 3W
Beachy Head England 13 **G2** 50N 0
Bedford England 21 **G4** 52N 0
Bedfordshire *d.* England 21 **G4** 52N 0
Beijing China 45 **G3** 40N 116E
Belfast N. Ire. 23 **F4** 54N 5W
Belgium Europe 33 **D3** 51N 4E
Belgrade Yugo. 33 **F2** 45N 21E
Belize Cen. America 49 **F1** 17N 89W
Belmopan Belize 49 **F1** 17N 89W
Bengal, B. of Indian Oc. 44 **E2** 17N 89E
Benin Africa 41 **B3** 9N 2E
Ben Lawers *mtn.* Scotland 13 **D5** 56N 4W
Ben Macdhui *mtn.* Scotland 13 **E5** 57N 3W
Ben More *mtn.* Scotland 13 **C5** 56N 6W
Ben Nevis *mtn.* Scotland 13 **D5** 56N 5W
Ben Wyvis *mtn.* Scotland 13 **D5** 57N 4W
Bering Sea N. America/Asia 44 **J4** 60N 175E
Bering Str. U.S.S.R./U.S.A. 48 **B4** 65N 170W
Berkshire *d.* England 21 **F3** 51N 1W
Berlin E. Germany 33 **E3** 53N 13E
Bermuda *i.* Atlantic Oc. 48 **G2** 32N 65W
Berne Switz. 33 **D2** 47N 7E
Berwick-upon-Tweed England 20 **E7** 55N 2W
Beverley England 20 **G5** 53N 0
Bhutan Asia 45 **F3** 27N 90E
Bill of Portland *c.* England 13 **E2** 50N 2W
Birkenhead England 20 **D5** 53N 3W
Birmingham England 21 **F4** 52N 1W
Biscay, B. of France 32 **C2** 45N 4W
Bishop Auckland England 20 **F6** 54N 1W
Bissau Guinea Bissau 41 **A3** 12N 16W
Blackburn England 20 **E5** 53N 2W
Blackpool England 20 **D5** 53N 3W
Black Sea Europe 32 **G2** 44N 35E
Black Volta *r.* Ghana/Burkina 40 **A3** 8N 2W
Blackwater *r.* Rep. of Ire. 13 **B3** 52N 8W
Blanc, Mont *mtn.* Europe 32 **D2** 46N 7E
Blue Mts. Australia 46 **E2** 34S 150E
Bodmin Moor England 13 **D2** 50N 4W
Bogotá Colombia 51 **B4** 5N 74W
Bolivia S. America 51 **B3** 17S 65W
Bolton England 20 **E5** 53N 2W
Bonn W. Germany 33 **D3** 51N 7E
Borders *d.* Scotland 22 **F2** 55N 2W
Borneo *i.* Asia 44 **G2** 1N 114E
Boston England 20 **G4** 52N 0
Boston U.S.A. 49 **G3** 42N 71W
Botany B. Australia 46 **E2** 34S 151E
Bothnia, G. of Europe 32 **E4** 63N 20E
Botswana Africa 41 **C1** 22S 24E
Bougainville *i.* P.N.G. 46 **E5** 6S 155E
Bournemouth England 21 **F2** 50N 1W
Boyne *r.* Rep. of Ire. 13 **C3** 53N 6W
Bradford England 20 **F5** 53N 1W
Brasília Brazil 51 **C3** 16S 48W
Brazil S. America 51 **C3** 10S 52W
Brazzaville Congo 41 **B2** 4S 15E
Brecon Wales 21 **D3** 51N 3W
Brecon Beacons *mts.* Wales 13 **E2** 51N 3W
Bressay *i.* Scotland 13 **H7** 60N 1W
Bridgwater England 21 **D3** 51N 3W
Bridlington England 20 **G6** 54N 0
Brighton England 21 **G2** 50N 0
Bristol England 21 **E3** 51N 2W
Bristol Channel U.K. 13 **E2** 51N 3W
British Isles Europe 32 **C3** 54N 3W
Brunei Asia 45 **G2** 5N 115E
Brussels Belgium 33 **D3** 51N 4E
Bucharest Romania 33 **F2** 44N 26E
Buckie Scotland 22 **F4** 57N 2W
Buckinghamshire *d.* England 21 **G3** 51N 0
Budapest Hungary 33 **E2** 47N 19E
Buenos Aires Argentina 51 **C2** 35S 59W
Bulgaria Europe 33 **F2** 42N 25E
Burkina Africa 41 **A3** 12N 2W
Burma Asia 45 **F3** 21N 96E
Burnley England 20 **E5** 53N 2W
Burton-upon-Trent England 21 **F4** 52N 1W
Burundi Africa 41 **C2** 3S 30E
Bury St. Edmunds England 21 **H4** 52N 0
Butt of Lewis *c.* Scotland 13 **C6** 58N 6W

Caernarfon Wales 20 **C5** 53N 4W
Caernarfon B. Wales 13 **D3** 53N 4W
Caerphilly Wales 21 **D3** 51N 3W
Cairngorms *mts.* Scotland 13 **E5** 57N 3W
Cairo Egypt 41 **C4** 30N 31E
Calgary Canada 49 **E3** 51N 114W
Callander Scotland 22 **D3** 56N 4W
Cambodia Asia 45 **F2** 12N 105E
Cambrian Mts. Wales 13 **E3** 52N 3W
Cambridge England 21 **H4** 52N 0
Cambridgeshire *d.* England 21 **G4** 52N 0
Cameroon Africa 41 **B3** 6N 13E
Campbeltown Scotland 22 **C2** 55N 5W
Canada N. America 49 **E3** 60N 105W
Canadian Shield *f.* N. America 48 **F3** 50N 80W
Canary Is. Atlantic Oc. 40 **A4** 29N 15W
Canberra Australia 47 **D2** 35S 149E
Cantabrian Mts. Spain 32 **C2** 43N 5W
Canterbury England 21 **I3** 51N 1E
Cape Breton I. Canada 48 **G3** 46N 61W
Cape Town R.S.A. 41 **B1** 34S 18E
Caracas Venezuela 51 **B4** 11N 67W
Cardiff Wales 21 **D3** 51N 3W
Cardigan Wales 21 **C4** 52N 4W
Cardigan B. Wales 13 **D3** 52N 4W
Caribbean Sea Cen. America 48 **G1** 15N 75W
Carlisle England 20 **E6** 54N 2W
Carlow *d.* Rep. of Ire. 23 **E2** 52N 6W
Carmarthen Wales 21 **C3** 51N 4W
Carmarthen B. Wales 13 **D2** 51N 4W
Carnsore Pt. Rep. of Ire. 13 **C3** 52N 6W
Carpathian Mts. Europe 32 **F2** 49N 24E
Carpentaria, G. of Australia 46 **C4** 14S 140E
Carrauntoohil *mtn.* Rep. of Ire. 13 **B2** 51N 9W
Carrickfergus N. Ire. 23 **F4** 54N 5W
Caspian Sea U.S.S.R. 44 **D4** 42N 51E
Castle Douglas Scotland 22 **E1** 54N 3W
Castleford England 20 **F5** 53N 1W
Caucasus Mts. U.S.S.R. 32 **H2** 43N 44E
Cavan *d.* Rep. of Ire. 23 **D3** 53N 7W
Cayenne Guiana 51 **C4** 5N 52W
Celebes *i.* Indonesia 44 **G1** 2S 120E
Central *d.* Scotland 22 **D3** 56N 4W
Central African Republic Africa 41 **B3** 6N 20E
Central Russian Uplands *f.* U.S.S.R. 32 **G3** 53N 35E
Central Siberian Upland *f.* U.S.S.R. 44 **F5** 66N 108E
Chad Africa 41 **B3** 13N 19E
Chad, L. Africa 40 **B3** 13N 14E
Chang Jiang *r.* China 44 **G3** 32N 121E
Channel Islands *d.* U.K. 21 **E1** 49N 2W
Chatham England 21 **H3** 51N 0
Chelmsford England 21 **H3** 51N 0
Cheltenham England 21 **E3** 51N 2W
Cheshire *d.* England 20 **E5** 53N 2W
Chester England 20 **E5** 53N 2W
Chesterfield England 20 **F5** 53N 1W
Chicago U.S.A. 49 **F3** 42N 88W
Chichester England 21 **G2** 50N 0
Chile S. America 51 **B2** 33S 71W
Chiltern Hills England 13 **F2** 51N 0
China Asia 45 **F3** 33N 103E
Christchurch England 21 **F2** 50N 1W
Clare *d.* Rep. of Ire. 23 **C2** 52N 8W
Clear, C. Rep. of Ire. 13 **B2** 51N 9W
Cleethorpes England 20 **G5** 53N 0
Cleveland *d.* England 20 **F6** 54N 1W
Cleveland Hills England 13 **F4** 54N 1W
Clwyd *d.* Wales 20 **D5** 53N 3W
Clyde *r.* Scotland 13 **D4** 55N 4W
Coatbridge Scotland 22 **D2** 55N 4W
Colchester England 21 **H3** 51N 0
Coleraine N. Ire. 23 **E5** 55N 6W
Coll *i.* Scotland 13 **C5** 56N 6W
Colombia S. America 51 **B4** 5N 75W
Colombo Sri Lanka 45 **E2** 7N 80E
Colorado *r.* U.S.A. 48 **E2** 28N 96W
Columbia *r.* U.S.A. 48 **D3** 46N 123W
Colwyn Wales 20 **D5** 53N 3W
Comoros Africa 41 **D2** 12S 44E
Conakry Guinea 41 **A3** 10N 14W
Congo Africa 41 **B2** 1S 16E
Conn, Lough Rep. of Ire. 13 **B4** 54N 9W
Cook, Mt. New Zealand 46 **G1** 44S 170E
Cook Str. New Zealand 46 **G1** 41S 174E
Cooper Creek *r.* Australia 46 **C3** 29S 138E
Copenhagen Denmark 33 **E3** 56N 12E
Coral Sea Pacific Oc. 46 **E4** 16S 155E
Corby England 21 **G4** 52N 0
Cork *d.* Rep. of Ire. 23 **C1** 51N 8W
Cork Rep. of Ire. 23 **C1** 51N 8W
Cornwall *d.* England 21 **C2** 50N 4W
Corrib, Lough Rep. of Ire. 13 **B3** 53N 9W
Corsica *i.* France 32 **D2** 42N 9E
Costa Rica Cen. America 49 **F1** 10N 84W
Cotswold Hills England 13 **E2** 51N 2W
Coventry England 21 **F4** 52N 1W
Craigavon N. Ire. 23 **E4** 54N 6W
Crawley England 21 **G3** 51N 0
Crete *i.* Greece 32 **F1** 35N 25E
Crewe England 20 **E5** 53N 2W
Crieff Scotland 22 **E3** 56N 3W
Cross Fell *mtn.* England 13 **E4** 54N 2W
Cuba Cen. America 49 **G2** 22N 79W
Cuillin Hills Scotland 13 **C5** 57N 6W
Cumbernauld Scotland 22 **D2** 55N 4W

umbria d. England 20 D6 54N 3W
umbrian Mts. England 13 E4 54N 3W
umnock Scotland 22 D2 55N 4W
upar Scotland 22 E3 56N 3W
uraçao i. Neth. Antilles 50 B4 12N 69W
wmbran Wales 21 D3 51N 3W
yprus Asia 45 C3 35N 33E
zechoslovakia Europe 33 E2 49N 15E

akar Senegal 41 A3 15N 17W
alkeith Scotland 22 E2 55N 3W
allas U.S.A. 49 F2 33N 97W
amascus Syria 45 C3 33N 36E
anube r. Europe 32 F2 45N 29E
arling r. Australia 46 D2 34S 142E
arling Downs f. Australia 46 D3 28S 150E
arling Range mts. Australia 46 A2 32S 116E
arlington England 20 F6 54N 1W
artmoor hills. England 13 E2 50N 3W
avis Str. N. America 48 H4 66N 58W
ead Sea Jordan 44 C3 31N 35E
eal England 21 I3 51N 1E
eccan f. India 44 E2 18N 77E
ee r. Scotland 13 E5 57N 2W
ee r. Wales 13 E3 53N 3W
enbigh Wales 20 D5 53N 3W
enmark Europe 33 D3 55N 10E
enver U.S.A. 49 F2 39N 105W
erby England 21 F4 52N 1W
erbyshire d. England 20 F5 53N 1W
erg, Lough Rep. of Ire. 13 B3 52N 8W
etroit U.S.A. 49 F3 42N 83W
evon d. England 21 D2 50N 3W
haka Bangla. 45 F3 24N 90E
inaric Alps mts. Yugo. 32 E2 44N 16E
ingle B. Rep. of Ire. 13 A3 52N 10W
ingwall Scotland 22 D4 57N 4W
isappointment, L. Australia 46 B3 24S 123E
jibouti Africa 41 D3 12N 43E
nepropetrovsk U.S.S.R. 33 G2 49N 35E
nestr r. U.S.S.R. 32 F2 46N 30E
nieper r. U.S.S.R. 32 G2 47N 32E
odoma Tanzania 41 C2 6S 36E
olgellau Wales 21 D4 52N 3W
ominica Windward Is. 49 G1 15N 61W
ominican Republic Cen. America 49 G1 18N 70W
on r. Scotland 13 E5 57N 2W
on r. U.S.S.R. 32 G2 47N 39E
oncaster England 20 F5 53N 1W
onegal d. Rep. of Ire. 23 C4 54N 8W
onegal B. Rep. of Ire. 13 B4 54N 8W
onets r. U.S.S.R. 32 H2 48N 41E
orchester England 21 E2 50N 2W
ornoch Firth est. Scotland 13 D5 57N 4W
orset d. England 21 E2 50N 2W
ouglas I.O.M. 20 C6 54N 4W
ouro r. Portugal 32 C2 41N 9W
over England 21 I3 51N 1E
own d. N. Ire. 23 E4 54N 6W
ownpatrick N. Ire. 23 F4 54N 5W
rakensberg mts. R.S.A. 40 C1 30S 29E
ublin d. Rep. of Ire. 23 E3 53N 6W
ublin Rep. of Ire. 23 E3 53N 6W
udley England 21 E4 52N 2W
umbarton Scotland 22 D2 55N 4W
umfries Scotland 22 E2 55N 3W
umfries and Galloway d. Scotland 22 E2 55N 3W
undalk Rep. of Ire. 23 E4 54N 6W
undalk B. Rep. of Ire. 13 C3 53N 6W
undee Scotland 22 F3 56N 2W
unfermline Scotland 22 E3 56N 3W
ungannon N. Ire. 23 E4 54N 6W
ungeness c. England 13 G2 50N 0
un Laoghaire Rep. of Ire. 23 E3 53N 6W
unoon Scotland 22 D2 55N 4W
uns Scotland 22 F2 55N 2W
urham England 20 F6 54N 1W
urham d. England 20 F6 54N 1W
urness Scotland 22 D5 58N 4W
vina r. U.S.S.R. 32 F3 57N 24E
vina, North r. U.S.S.R. 32 H4 65N 41E
yfed d. Wales 21 C4 52N 4W

arn r. Scotland 13 E5 56N 3W
astbourne England 21 H2 50N 0
astern Ghats mts. India 44 E2 16N 80E
ast Germany Europe 33 E3 52N 12E
ast Kilbride Scotland 22 D2 55N 4W
astleigh England 21 F2 50N 1W
ast Sussex d. England 21 H2 50N 0

Ebro r. Spain 32 D2 41N 1E
Ecuador S. America 51 B3 2S 78W
Edinburgh Scotland 22 E2 55N 3W
Egypt Africa 41 C4 26N 30E
El Aaiún W. Sahara 41 A4 27N 13W
Elbe r. West Germany 32 D3 54N 10E
Elbrus mtn. U.S.S.R. 44 C4 43N 42E
Elgin Scotland 22 E4 57N 3W
El Salvador Cen. America 49 F1 14N 89W
England d. U.K. 21 F4 52N 1W
English Channel U.K. 32 C2 50N 1W
Enniskillen N. Ire. 23 D4 54N 7W
Epsom England 21 G3 51N 0
Equatorial Guinea Africa 41 B3 2N 10E
Erie, L. Canada/U.S.A. 48 F3 42N 81W
Erris Head Rep. of Ire. 13 A4 54N 10W
Esher England 21 G3 51N 0
Essex d. England 21 H3 51N 0
Ethiopan Highlands Ethiopa 40 C3 10N 37E
Ethiopia Africa 41 C3 10N 39E
Etna, Mt. Italy 32 E1 38N 15E
Euphrates r. Asia 44 C3 31N 47E
Everest, Mt. Asia 44 E3 28N 87E
Exe r. England 13 E2 50N 3W
Exeter England 21 D2 50N 3W
Exmoor England 13 E2 51N 3W
Exmouth England 21 D2 50N 3W
Eyemouth Scotland 22 F2 55N 2W
Eyre, L. Australia 46 C3 28S 137E

Fair Isle Scotland 13 H6 59N 1W
Falkirk Scotland 22 E3 56N 3W
Falkland Is. S. America 50 C1 52S 60W
Falmouth England 21 B2 50N 5W
Fareham England 21 F2 50N 1W
Farnborough England 21 G3 51N 0
Farnham England 21 G3 51N 0
Faroe Is. Europe 32 C4 62N 7W
Fermanagh d. N. Ire. 23 D4 54N 7W
Fife d. Scotland 22 E3 56N 3W
Fiji Pacific Oc. 55 I3 18S 178E
Finisterre, C. Spain 32 C2 43N 9W
Finland Europe 33 F4 64N 27E
Firth of Clyde est. Scotland 13 D4 55N 4W
Firth of Forth est. Scotland 13 E5 56N 3W
Firth of Lorn est. Scotland 13 D5 56N 5W
Firth of Tay est. Scotland 13 E5 56N 3W
Fishguard Wales 21 C3 51N 4W
Flamborough Head England 13 F4 54N 0
Fleetwood England 20 D5 53N 3W
Flinders r. Australia 46 D4 17S 141E
Flinders Range mts. Australia 46 C2 31S 139E
Florida f. U.S.A. 48 F2 29N 82W
Folkestone England 21 I3 51N 1E
Forfar Scotland 22 F3 56N 3W
Forres Scotland 22 E4 57N 3W
Forth r. Scotland 13 E5 56N 3W
Fort William Scotland 22 C3 56N 5W
Foula i. Scotland 13 G7 60N 2W
Foyle, Lough Rep. of Ire./N. Ire. 13 C4 55N 7W
France Europe 33 D2 47N 2E
Fraserburgh Scotland 22 F4 57N 2W
Freetown Sierra Leone 41 A3 9N 13W
Futa Jalon Plateau f. Guinea 40 A3 11N 12W

Gabon Africa 41 B2 0 12E
Gaborone Botswana 41 C1 25S 26E
Galapagos Is. Pacific Oc. 50 A4 0 90W
Galashiels Scotland 22 F2 55N 2W
Gallinas, C. Colombia 50 B4 12N 72W
Galty Mts. Rep. of Ire. 13 B3 52N 8W
Galway d. Rep. of Ire. 23 B3 53N 9W
Galway Rep. of Ire. 23 B3 53N 9W
Galway B. Rep. of Ire. 13 B3 53N 9W
Gambia Africa 41 A3 13N 16W
Gambia r. Gambia 40 A3 13N 16W
Ganges r. India 44 E3 23N 90E
Garonne r. France 32 C2 45N 1W
Georgetown Guyana 51 C4 7N 58W
Ghana Africa 41 A3 8N 1W
Gibraltar Europe 33 C1 36N 5W
Gibraltar, Str. of Africa/Europe 32 C1 36N 5W
Gibson Desert Australia 46 B3 25S 125E
Gillingham England 21 H3 51N 0
Girvan Scotland 22 D2 55N 4W
Glasgow Scotland 22 D2 55N 4W
Glenrothes Scotland 22 E3 56N 3W
Gloucester England 21 E3 51N 2W
Gloucestershire d. England 21 E3 51N 2W
Goat Fell mtn. Scotland 13 D4 55N 5W
Gobi des. Asia 44 F4 43N 103E
Godthåb Greenland 49 H4 64N 52W
Golspie Scotland 22 E4 57N 3W

Gongga Shan mtn. China 44 F3 29N 101E
Good Hope, C. of R.S.A. 40 B1 34S 18E
Gorki U.S.S.R. 33/H3 56N 44E
Grampian d. Scotland 22 F4 57N 2W
Grampian Mts. Scotland 13 D5 56N 4W
Gran Chaco f. S. America 50 B2 24S 60W
Grangemouth Scotland 22 E3 56N 3W
Grantham England 21 G4 52N 0
Gravesend England 21 H3 51N 0
Grays England 21 H3 51N 0
Great Artesian Basin f. Australia 46 D3 26S 143E
Great Australian Bight Australia 46 B2 33S 130E
Great Barrier Reef f. Australia 46 D4 16S 146E
Great Bear L. Canada 48 E4 66N 120W
Great Dividing Range mts. Australia 46 D3 22S 148E
Greater Antilles is. Cen. America 48 G1 17N 70W
Greater London d. England 21 G3 51N 0
Greater Manchester d. England 20 E5 53N 2W
Great Malvern England 21 E4 52N 2W
Great Ouse r. England 13 G3 52N 0
Great Rift Valley f. Africa 40 C2 7S 33E
Great Salt L. U.S.A. 48 E3 41N 113W
Great Sandy Desert Australia 46 B3 22S 125E
Great Slave L. Canada 48 E4 61N 114W
Great Victoria Desert Australia 46 B3 29S 128E
Great Yarmouth England 21 I4 52N 1E
Greece Europe 33 F1 39N 22E
Greenland N. America 49 H4 68N 45W
Greenock Scotland 22 D2 55N 4W
Grimsby England 20 G5 53N 0
Gross Glockner mtn. Austria 32 E2 47N 13E
Guatemala Cen. America 49 F1 16N 90W
Guatemala City Guatemala 49 F1 15N 90W
Guiana S. America 51 C4 4N 53W
Guildford England 21 G3 51N 0
Guinea Africa 41 A3 10N 12W
Guinea Bissau Africa 41 A3 12N 15W
Guinea, G. of Africa 40 B3 3N 3E
Guyana S. America 51 C4 5N 59W
Gwent d. Wales 21 E3 51N 2W
Gwynedd d. Wales 20 C5 53N 4W

Haddington Scotland 22 F2 55N 2W
Hainan i. China 44 G2 19N 110E
Haiti Cen. America 49 G1 19N 73W
Halifax Canada 49 G3 45N 64W
Halifax England 20 F5 53N 1W
Hamburg W. Germany 33 D3 54N 10E
Hamersley Range mts. Australia 46 A3 22S 118E
Hamilton Scotland 22 D2 55N 4W
Hampshire d. England 21 F3 51N 1W
Hanoi Vietnam 45 F3 21N 106E
Harare Zimbabwe 41 C2 18S 31E
Harlow England 21 H3 51N 0
Harris i. Scotland 13 C5 57N 6W
Harrogate England 20 F5 53N 1W
Hartlepool England 20 F6 54N 1W
Harwich England 21 I3 51N 1E
Hastings England 21 H2 50N 0
Hatfield England 21 G3 51N 0
Hatteras, C. U.S.A. 48 G2 35N 75W
Havana Cuba 49 F2 23N 82W
Havant England 21 G2 50N 0
Haverfordwest Wales 21 C3 51N 4W
Hawaiian Is. U.S.A. 54 A5 21N 157W
Hawick Scotland 22 F2 55N 2W
Hekla Mt. Iceland 32 A4 64N 19W
Helsinki Finland 33 F4 60N 25E
Hemel Hempstead England 21 G3 51N 0
Hereford England 21 E4 52N 2W
Hereford and Worcester d. England 21 E4 52N 2W
Hertford England 21 G3 51N 0
Hertfordshire d. England 21 G3 51N 0
Highland d. Scotland 22 C4 57N 5W
High Peak mtn. England 13 F5 53N 1W
High Wycombe England 21 G3 51N 0
Himalaya mts. Asia 44 E3 29N 84E
Hokkaido i. Japan 44 H4 43N 144E
Holyhead Wales 20 C5 53N 4W
Holy I. England 13 F4 55N 1W
Honduras Cen. America 49 F1 14N 87W
Honiara Solomon Is. 47 E5 9S 160E
Honshu i. Japan 44 H3 36N 138E
Horn, C. S. America 50 B1 56S 67W
Hove England 21 G2 50N 0

Howe, C. Australia 46 D2 37S 150E
Hoy i. Scotland 13 E6 58N 3W
Huang He r. China 44 G3 38N 119E
Huddersfield England 20 F5 53N 1W
Hudson B. Canada 48 F3 58N 86W
Humber r. England 13 F3 53N 0
Humberside d. England 20 G5 53N 0
Hungarian Plain f. Hungary 32 E2 48N 19E
Hungary Europe 33 E2 47N 19E
Hunstanton England 21 H4 52N 0
Huntingdon England 21 G4 52N 0
Huron, L. Canada/U.S.A. 48 F3 45N 82W

Iceland Europe 33 B4 65N 18W
Ilfracombe England 21 C3 51N 4W
India Asia 45 E3 23N 78E
Indian Ocean 53 F3 5S 60E
Indonesia Asia 45 G1 6S 118E
Indus r. Pakistan 44 D3 24N 68E
Inner Hebrides is. Scotland 13 C5 56N 6W
Invergordon Scotland 22 D4 57N 4W
Inverness Scotland 22 D4 57N 4W
Inverurie Scotland 22 F4 57N 2W
Ipswich England 21 I4 52N 1E
Iran Asia 45 D3 32N 54E
Iranian Plateau f. Asia 44 D3 33N 55E
Iraq Asia 45 C3 33N 44E
Irish Sea U.K./Rep. of Ire. 13 D3 53N 5W
Irkutsk U.S.S.R. 45 F4 52N 104E
Irrawaddy r. Burma 44 F2 18N 95E
Irtysh r. U.S.S.R. 44 D5 61N 69E
Irvine Scotland 22 D2 55N 4W
Islamabad Pakistan 45 E3 34N 73E
Islay i. Scotland 13 C4 55N 6W
Isle of Man d. U.K. 20 C6 54N 4W
Isle of Wight d. England 21 F2 50N 1W
Israel Asia 45 C3 32N 34E
Italy Europe 33 E2 43N 12E
Ivory Coast Africa 41 A3 8N 5W

Jakarta Indonesia 45 F1 6S 107E
Jamaica Cen. America 49 G1 18N 77W
Jammu and Kashmir Asia 45 E3 34N 76E
Japan Asia 45 H3 36N 138E
Java i. Indonesia 44 F1 7S 110E
Jordan Asia 45 C3 31N 36E
Jura mts. France 32 D2 47N 7E
Jura i. Scotland 13 D4 55N 5W
Jutland f. Denmark 32 D3 56N 9E

Kabul Afghan. 45 D3 35N 69E
Kalahari Desert Botswana 40 C1 24S 23E
Kamchatka Pen. U.S.S.R. 44 I4 56N 160E
Kampala Uganda 41 C3 0 33E
Kariba, L. Zimbabwe/Zambia 40 C2 17S 28E
Kathmandu Nepal 45 E3 28N 85E
Kazan U.S.S.R. 33 H3 56N 49E
Keith Scotland 22 F4 57N 2W
Kelso Scotland 22 F2 55N 2W
Kendal England 20 E6 54N 2W
Kent d. England 21 H3 51N 0
Kenya Africa 41 C3 0 38E
Kerguelen i. Southern Oc. 55 F2 50S 70E
Kerry d. Rep. of Ire. 23 B2 52N 9W
Kettering England 21 G4 52N 0
Kharkov U.S.S.R. 33 G2 50N 36E
Khartoum Sudan 41 C3 16N 33E
Kiev U.S.S.R. 33 G3 51N 31E
Kildare d. Rep. of Ire. 23 E3 53N 6W
Kilimanjaro mtn. Tanzania 40 C2 3S 37E
Kilkenny d. Rep. of Ire. 23 D2 52N 7W
Kilmarnock Scotland 22 D2 55N 4W
Kinabalu mtn. Malaysia 44 G2 6N 117E
King Leopold Ranges mts. Australia 46 B4 17S 126E
Kings Lynn England 21 H4 52N 0
Kingston Jamaica 49 G1 18N 77W
Kingston upon Hull England 20 G5 53N 0
Kingussie Scotland 22 D4 57N 4W
Kinnaird's Head Scotland 13 F5 57N 2W
Kinross Scotland 22 E3 56N 3W
Kinshasa Zaïre 41 B2 4S 15E
Kirgiz Steppe f. U.S.S.R. 44 D4 49N 57E
Kirkcaldy Scotland 22 E3 56N 3W
Kirkcudbright Scotland 22 D1 54N 4W
Kirkwall Scotland 22 F5 58N 2W
Kosciusko, Mt. Australia 46 D2 36S 148E
Kuala Lumpur Malaysia 45 F2 3N 102E
Kuh-i-Hazar mtn. Iran 44 D3 30N 57E
Kunlun Shan mts. China 44 E3 37N 88E
Kuwait Asia 45 C3 29N 48E
Kuybyshev U.S.S.R. 33 I3 53N 50E

Kyle of Lochalsh town Scotland 22 C4 57N 5W
Kyushu i. Japan 44 H3 32N 130E

Labrador f. Canada 48 G3 54N 61W
Ladoga, L. U.S.S.R. 32 G4 61N 32E
Lairg Scotland 22 D5 58N 4W
Lake District f. England 13 E4 54N 3W
Lanark Scotland 22 E2 55N 4W
Lancashire d. England 20 E5 53N 2W
Lancaster England 20 E6 54N 2W
Land's End c. England 13 D2 50N 5W
Laois d. Rep. of Ire. 23 D2 52N 7W
Laos Asia 45 F2 19N 104E
La Paz Bolivia 51 B3 16S 68W
Lapland f. Sweden/Finland 32 F4 68N 24E
Largs Scotland 22 D2 55N 4W
Larne N. Ire. 23 E4 54N 5W
Lebanon Asia 45 C3 34N 36E
Lee r. Rep. of Ire. 13 B2 51N 8W
Leeds England 20 F5 53N 1W
Leeuwin, C. Australia 46 A2 34S 115E
Leicester England 21 F4 52N 1W
Leicestershire d. England 21 F4 52N 1W
Leitrim d. Rep. of Ire. 23 D4 54N 7W
Lena r. U.S.S.R. 44 G5 72N 127E
Leningrad U.S.S.R. 33 G3 60N 30E
Lerwick Scotland 22 G7 60N 1W
Lesotho Africa 41 C1 30S 28E
Lesser Antilles is. Cen. America 48 G1 13N 65W
Lewes England 21 H2 50N 0
Lewis i. Scotland 13 C6 58N 6W
Liberia Africa 41 A3 6N 10W
Libreville Gabon 41 B3 1N 9E
Libya Africa 41 B4 26N 17E
Libyan Desert Africa 40 C4 25N 26E
Lilongwe Malawi 41 C2 14S 34E
Lima Peru 51 B3 12S 77W
Limerick d. Rep. of Ire. 23 C2 52N 8W
Limerick Rep. of Ire. 23 C2 52N 8W
Limpopo r. Mozambique 40 C1 25S 34E
Lincoln England 20 G5 53N 0
Lincolnshire d. England 20 G5 53N 0
Lisbon Portugal 33 C1 39N 9W
Lisburn N. Ire. 23 E4 54N 6W
Liverpool England 20 D5 53N 3W
Livingston Scotland 22 E2 55N 3W
Llandrindod Wells Wales 21 D4 52N 3W
Llanos f. Colombia/Venezuela 50 B4 6N 72W
Lochboisdale town Scotland 22 A4 57N 7W
Lochgilphead Scotland 22 C3 56N 5W
Lochinver Scotland 22 C5 58N 5W
Lochmaddy town Scotland 22 A4 57N 7W
Lockerbie Scotland 22 E2 55N 3W
Lofoten Is. Norway 32 E4 68N 14E
Logan, Mt. Canada 48 D4 61N 140W
Loire r. France 32 C2 47N 2W
Lomé Togo 41 B3 6N 1E
Lomond, Loch Scotland 13 D5 56N 4W
London England 21 G3 51N 0
Londonderry N. Ire. 23 D5 55N 7W
Londonderry d. N. Ire. 23 D4 54N 7W
Longford d. Rep. of Ire. 23 D3 53N 7W
Lop Nur l. China 44 F4 40N 91E
Los Angeles U.S.A. 49 E2 34N 118W
Lossiemouth Scotland 22 E4 57N 3W
Lothian d. Scotland 22 E2 55N 3W
Louth d. Rep. of Ire. 23 E3 53N 6W
Lower Lough Erne N. Ire. 13 C4 54N 7W
Lowestoft England 21 I4 52N 1E
Luanda Angola 41 B2 9S 13E
Luce B. Scotland 13 D4 54N 4W
Lurgan N. Ire. 23 E4 54N 6W
Lusaka Zambia 41 C2 15S 28E
Luton England 21 G3 51N 0
Luxembourg Europe 33 D2 50N 6E
Luzon i. Phil. 44 G2 18N 121E
Lyme B. England 13 E2 50N 3W
Lyon France 33 D2 46N 5E

Macclesfield England 20 E5 53N 2W
Macdonnell Ranges mts. Australia 46 C3 23S 132E
Macgillycuddy's Reeks mts. Rep. of Ire. 13 B3 52N 9W
Mackenzie r. Canada 48 D4 69N 134W
Madagascar Africa 41 D2 20S 46E
Madeira i. Atlantic Oc. 40 A4 33N 17W
Madrid Spain 33 C2 40N 4W
Magellan, Str. of Chile 50 B1 53S 71W
Maidenhead England 21 G3 51N 0
Maidstone England 21 H3 51N 0

Mainland, Orkney i. Scotland 13 E6 59N 3W
Mainland, Shetland i. Scotland 13 H7 60N 1W
Makran f. Asia 44 D3 26N 60E
Malawi Africa 41 C2 13S 34E
Malawi, L. Africa 40 C2 12S 35E
Malaysia Asia 45 F2 5N 110E
Maldives Indian Oc. 45 E2 6N 73E
Mali Africa 41 A3 18N 2W
Malin Head Rep. of Ire. 13 C4 55N 7W
Mallaig Scotland 22 C4 57N 5W
Malta Europe 33 E1 36N 14E
Managua Nicaragua 49 F1 12N 86W
Manchester England 20 E5 53N 2W
Manchurian Plain f. Asia 44 G4 42N 122E
Manila Phil. 45 G2 15N 121E
Man, Isle of U.K. 13 D4 54N 4W
Mansfield England 20 F5 53N 1W
Maputo Mozambique 41 C1 26S 33E
Maracaibo, L. Venezuela 50 B4 10N 72W
Maree, Loch Scotland 13 D5 58N 5W
Margate England 21 I3 51N 1E
Marseille France 33 D2 43N 5E
Marshall Is. Pacific Oc. 55 I4 10N 172E
Maseru Lesotho 41 C1 29S 27E
Mask, Lough Rep. of Ire. 13 B3 53N 9W
Massif Central mts. France 32 D2 45N 3E
Matlock England 20 F5 53N 1W
Mauritania Africa 41 A4 20N 10W
Mayo d. Rep. of Ire. 23 B3 53N 9W
Mbabane Swaziland 41 C1 26S 31E
McKinley, Mt. U.S.A. 48 C4 63N 151W
Meath d. Rep. of Ire. 23 E3 53N 6W
Mediterranean Sea 32 E1 37N 15E
Mekong r. Asia 44 F2 10N 106E
Mendip Hills England 13 E2 51N 2W
Mersey r. England 13 E3 53N 2W
Merseyside d. England 20 E5 53N 2W
Merthyr Tydfil Wales 21 D3 51N 3W
Meseta f. Spain 32 C2 40N 4W
Mexico Cen. America 49 E2 20N 100W
Mexico City Mexico 49 F1 19N 99W
Mexico, G. of N. America 48 F2 25N 90W
Miami U.S.A. 49 F2 26N 80W
Michigan, L. U.S.A. 48 F3 44N 87W
Middlesbrough England 20 F6 54N 1W
Mid Glamorgan d. Wales 21 D3 51N 3W
Milan Italy 32 D2 45N 9E
Milford Haven Wales 21 B3 51N 5W
Milton Keynes England 21 G4 52N 0
Mindanao i. Phil. 44 G2 7N 125E
Minehead England 13 D3 51N 3W
Minsk U.S.S.R. 33 F3 54N 27E
Mississippi r. U.S.A. 48 F2 29N 89W
Missouri r. U.S.A. 48 F2 39N 90W
Mogadishu Somali Rep. 41 D3 2N 45E
Mold Wales 20 D5 53N 3W
Monaco Europe 33 D2 44N 7E
Monaghan d. Rep. of Ire. 23 D4 54N 7W
Mongolia Asia 45 F4 46N 104E
Monrovia Liberia 41 A3 6N 11W
Montevideo Uruguay 51 C2 35S 56W
Montreal Canada 49 G3 45N 74W
Montrose Scotland 22 F3 56N 2W
Moray Firth est. Scotland 13 D5 57N 4W
Morcambe B. England 13 E4 54N 3W
Morocco Africa 41 A4 31N 5W
Morpeth England 20 F7 55N 1W
Moscow U.S.S.R. 33 G3 56N 38E
Motherwell Scotland 22 E2 55N 3W
Mourne r. N. Ire. 13 C4 54N 7W
Mourne Mts. N. Ire. 13 C4 54N 6W
Mozambique Africa 41 C2 18S 35E
Mozambique Channel Indian Oc. 40 D2 16S 42E
Mull i. Scotland 13 C5 56N 6W
Munich W. Germany 33 E2 48N 12E
Murray r. Australia 46 C2 35S 139E
Murrumbidgee r. Australia 46 D2 35S 143E
Muscat Oman 45 D3 24N 59E
Musgrave Ranges mts. Australia 46 C3 27S 131E
Musselburgh Scotland 22 E2 55N 3W

Nairn Scotland 22 E4 57N 3W
Nairobi Kenya 41 C2 1S 37E
Namibia Africa 41 B1 22S 17E
Naples Italy 33 E2 41N 14E
Nasser, L. Egypt 40 C4 23N 32E
N'Djamena Chad 41 B3 12N 15E
Neagh, Lough N. Ire. 13 C4 54N 6W
Nene r. England 13 F3 52N 0
Nepal Asia 45 E3 28N 84E

Ness, Loch Scotland 13 D5 57N 4W
Netherlands Europe 33 D3 52N 5E
Nevada, Sierra mts. Spain 32 C1 37N 3W
New Britain i. P.N.G. 46 E5 6S 150E
Newbury England 21 F3 51N 1W
New Caledonia i. Pacific Oc. 47 F3 22S 165E
Newcastle N. Ire. 23 F4 54N 5W
Newcastle upon Tyne England 20 F6 54N 1W
New Delhi India 45 E3 29N 77E
New Forest f. England 13 F2 50N 1W
Newfoundland i. Canada 48 H3 48N 56W
New Guinea i. Austa. 46 D5 5S 140E
New Hebrides is. Pacific Oc. 46 F4 16S 167E
New Orleans U.S.A. 49 F2 30N 90W
Newport England 21 F2 50N 1W
Newport Wales 21 D3 51N 3W
Newquay England 21 B2 50N 5W
Newry N. Ire. 23 E4 54N 6W
Newton Aycliffe England 20 F6 54N 1W
Newton Stewart Scotland 22 D1 54N 4W
Newtown Wales 21 D4 52N 3W
Newtownabbey N. Ire. 23 F4 54N 5W
Newtownards N. Ire. 23 F4 54N 5W
Newtown St. Boswells Scotland 22 F2 55N 2W
New York U.S.A. 49 G3 41N 74W
New Zealand Austa. 47 G1 41S 175E
Niamey Niger 41 B3 14N 2E
Nicaragua Cen. America 49 F1 13N 85W
Nicaragua, L. Nicaragua 48 F1 12N 85W
Nicobar Is. India 44 F2 8N 94E
Niger Africa 41 B3 17N 10E
Niger r. Nigeria 40 B3 4N 6E
Nigeria Africa 41 B3 9N 9E
Nile r. Egypt 40 C4 32N 30E
Nith r. Scotland 13 E4 55N 3W
Nore r. Rep. of Ire. 13 C3 52N 7W
Norfolk d. England 21 H4 52N 0
Norfolk Broads f. England 13 G3 52N 1E
Northallerton England 20 F6 54N 1W
Northampton England 21 G4 52N 0
Northamptonshire d. England 21 G4 52N 0
North Berwick Scotland 22 F3 56N 2W
North Channel U.K. 13 D4 55N 5W
North China Plain f. China 44 G3 34N 117E
North Downs hills England 13 G2 51N 0
Northern Ireland d. U.K. 23 D4 54N 7W
North European Plain f. Europe 32 F3 56N 27E
North I. New Zealand 46 G2 39S 175E
North Korea Asia 45 G4 40N 128E
North Sea Europe 32 D3 56N 3E
North Uist i. Scotland 13 C5 57N 7W
Northumberland d. England 20 E7 55N 2W
North West Highlands Scotland 13 D5 57N 5W
North York Moors hills England 13 F4 54N 0
North Yorkshire d. England 20 F6 54N 1W
Norway Europe 33 E4 65N 13E
Norwich England 21 I4 52N 1E
Nottingham England 21 F4 52N 1W
Nouakchott Mauritania 41 A3 18N 16W
Nouméa New Cal. 47 F3 22S 166E
Novosibirsk U.S.S.R. 45 E4 55N 83E
Nullarbor Plain f. Australia 46 B2 31S 128E

Ob r. U.S.S.R. 44 D5 67N 69E
Oban Scotland 22 C3 56N 5W
Ochil Hills Scotland 13 E5 56N 3W
Oder r. Poland 32 E3 53N 15E
Odessa U.S.S.R. 33 G2 47N 31E
Offaly d. Rep. of Ire. 23 D3 53N 7W
Okhotsk, Sea of Asia 44 H4 55N 150E
Oldham England 20 E5 53N 2W
Olympus, Mt. Greece 32 F2 40N 22E
Omagh N. Ire. 23 D4 54N 7W
Oman Asia 45 D3 22N 57E
Omsk U.S.S.R. 45 E4 55N 73E
Onega, L. U.S.S.R. 32 G4 62N 35E
Ontario, L. N. America 48 G3 44N 78W
Orange r. R.S.A. 40 B1 29S 16E
Orinoco r. Venezuela 50 B4 9N 61W
Orkney d. Scotland 22 F5 59N 3W
Orkney Is. Scotland 13 E6 59N 3W
Ormskirk England 20 E5 53N 2W
Oslo Norway 33 E3 60N 11E
Ottawa Canada 49 G3 45N 76W
Ouagadougou Burkina 41 A3 12N 2W
Ouse r. England 13 F4 54N 1W
Outer Hebrides is. Scotland 13 C5 57N 7W
Owen Stanley Range mts. P.N.G. 46 D5 9S 148E
Oxford England 21 F3 51N 1W
Oxfordshire d. England 21 F3 51N 1W
Ozark Plateau U.S.A. 48 F2 36N 94W

Pacific Ocean 52-53 A4 5N 150E
Padstow England 21 C2 50N 4W
Paisley Scotland 22 D2 55N 4W
Pakistan Asia 45 D3 30N 70E
Palermo Italy 33 E1 38N 13E
Pampas f. Argentina 50 B2 35S 63W
Panama Cen. America 49 F1 9N 80W
Panama City Panama 49 G1 9N 79W
Panama, Isthmus of Cen. America 48 F1 9N 83W
Papua New Guinea Austa. 47 D5 6S 143E
Paraguay S. America 51 C2 23S 58W
Paramaribo Surinam 51 C4 6N 55W
Paraná r. Argentina 50 C2 34S 59W
Paris France 33 D2 49N 2E
Parrett r. England 13 E2 51N 2W
Patagonia f. Argentina 50 B1 45S 68W
Peace r. Canada 48 E3 59N 111W
Pechora r. U.S.S.R. 32 I4 68N 54E
Peebles Scotland 22 E2 55N 3W
Peipus, L. U.S.S.R. 32 F3 59N 28E
Pembroke Wales 21 C3 51N 4W
Penrith England 20 E6 54N 2W
Pentland Firth Scotland 13 E6 58N 3W
Penzance England 21 B2 50N 5W
Perm U.S.S.R. 33 I3 58N 56E
Perth Scotland 22 E3 56N 3W
Peru S. America 51 B3 10S 75W
Peterborough England 21 G4 52N 0
Peterhead Scotland 22 G4 57N 1W
Peterlee England 20 F6 54N 1W
Philadelphia U.S.A. 49 G2 40N 75W
Philippines Asia 45 G2 13N 123E
Phnom Penh Cambodia 45 F2 12N 105E
Pindus Mts. Albania/Greece 32 F2 40N 21E
Pittsburgh U.S.A. 49 F3 40N 80W
Plymouth England 21 C2 50N 4W
Po r. Italy 32 E2 45N 12E
Poland Europe 33 E3 53N 19E
Pontypool Wales 21 D3 51N 3W
Poole England 21 F2 50N 1W
Portadown N. Ire. 23 E4 54N 6W
Portaferry N. Ire. 23 F4 54N 5W
Port-au-Prince Haiti 49 G1 19N 72W
Port Moresby P.N.G. 47 D5 9S 147E
Porto Novo Benin 41 B3 7N 3E
Portree Scotland 22 B4 57N 6W
Portrush N. Ire. 23 E5 55N 6W
Portsmouth England 21 F2 50N 1W
Port Talbot Wales 21 D3 51N 3W
Portugal Europe 33 C1 39N 8W
Powys d. Wales 21 D4 52N 3W
Prague Czech. 33 E2 50N 14E
Preston England 20 E5 53N 2W
Prestwick Scotland 22 D2 55N 4W
Pretoria R.S.A. 41 C1 26S 28E
Puerto Rico Cen. America 49 G1 18N 67W
Pyongyang N. Korea 45 G3 39N 126E
Pyrenees mts. France/Spain 32 D2 43N 1E

Qatar Asia 45 D3 25N 51E
Quebec Canada 49 G3 47N 71W
Quito Ecuador 51 B4 0 79W

Rabat Morocco 41 A4 34N 7W
Ramsgate England 21 I3 51N 1E
Rangoon Burma 45 F2 17N 96E
Rathlin I. N. Ire. 13 C4 55N 6W
Reading England 21 G3 51N 0
Red Basin f. China 44 F3 30N 105E
Redditch England 21 F4 52N 1W
Red Sea Africa/Asia 40 C4 20N 39E
Ree, Lough Rep. of Ire. 13 C3 53N 7W
Reigate England 21 G3 51N 0
Republic of Ireland Europe 23 D3 53N 7W
Republic of South Africa Africa 41 C1 30S 27E
Reykjavik Iceland 33 A4 64N 22W
Rhine r. Europe 33 D3 52N 6E
Rhodes i. Greece 32 F1 36N 28E
Rhondda Wales 21 D3 51N 3W
Rhône r. France 32 D2 43N 5E
Rhum i. Scotland 13 C5 57N 6W
Ribble r. England 13 E3 53N 2W
Rio de la Plata est. Uruguay/Argentina 50 C2 35S 56W
Rio Grande r. N. America 48 F2 26N 97W
Riyadh Saudi Arabia 45 C3 25N 47E
Rochdale England 20 E5 53N 2W
Rochester England 21 H3 51N 0
Rocky Mts. N. America 48 E3 42N 110W
Romania Europe 33 F2 46N 24E
Rome Italy 33 E2 42N 12E